MW01362706

# Science Activity Books
# MARINE BIOLOGY

Created by The Good and the Beautiful Team

Cover Illustrated by Sandra Eide
Pages Illustrated by Becca Hall
Design by Phillip Colhouer

© 2024 The Good and the Beautiful, LLC
goodandbeautiful.com

Circle the fish that matches the first fish in each row. There may be more than one matching fish in each row. Color the fish.

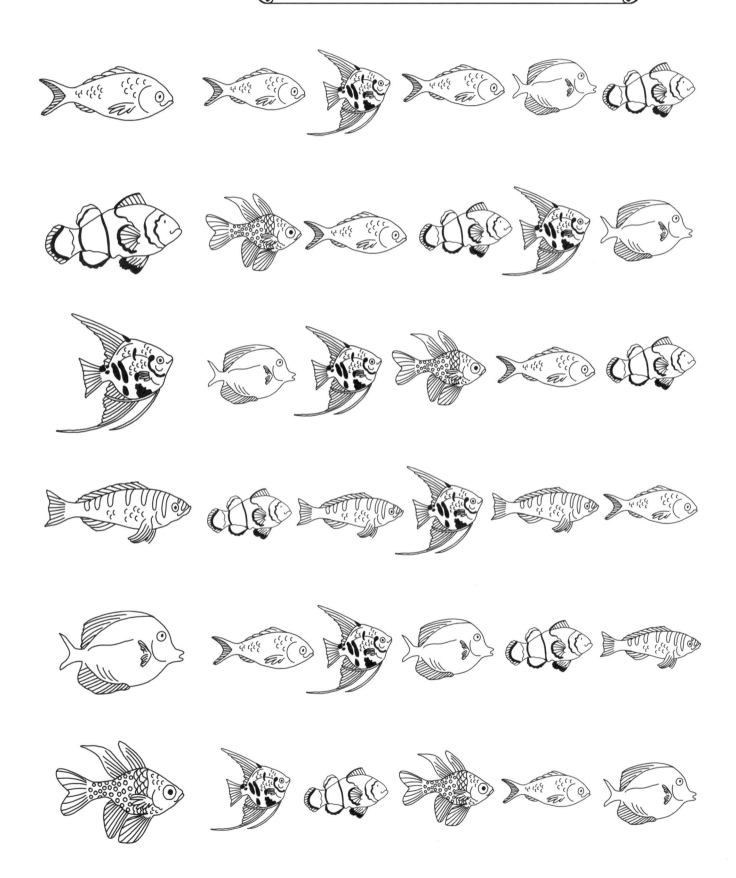

Lesson 2

Color each sea creature. Then draw a line from the sea creature to its matching shadow.

Trace the lines to connect each sea creature to its baby. Color the sea creatures.

**Lesson 3**

Circle the six items in the top picture that you do not see in the bottom picture. Color the top picture.

Lesson 4

Cut out the puzzle pieces. Arrange the puzzle correctly. Color the pieces. Tape or paste the puzzle pieces together on the top half of this page.

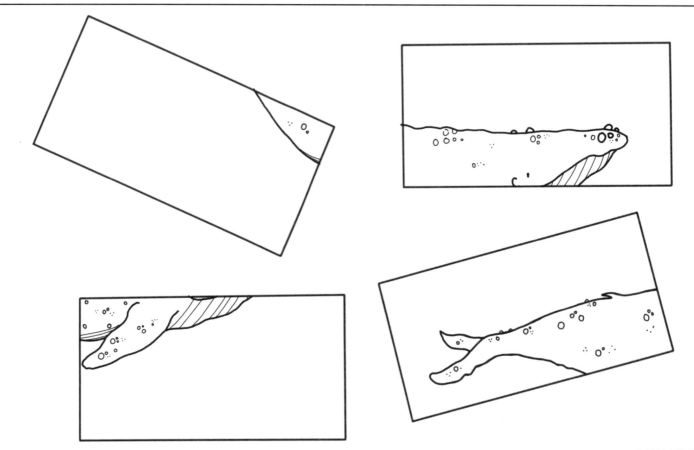

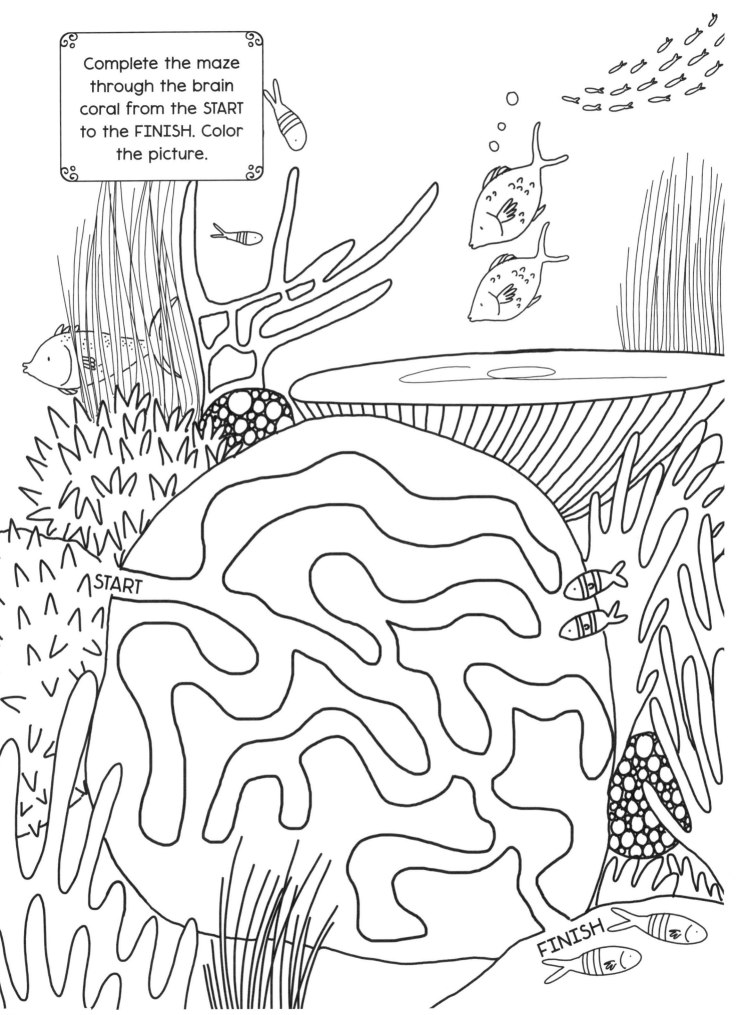

Color the number of sea creatures listed at the beginning of each row.

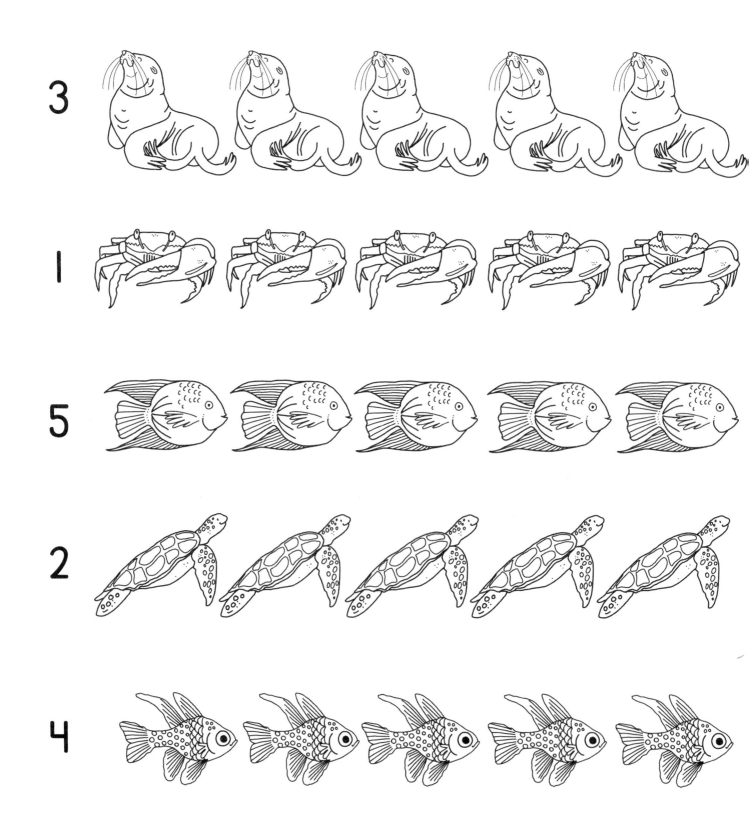

Find and color the parts that belong to the fish. Color the fish.

Draw the missing tentacles on the jellyfish. Color the picture.

Connect the dots.
Color the sea turtle.

Color the pictures. Cut out the pictures. Tape or paste the pictures in the correct boxes.

| Sea Creatures | Not Sea Creatures |
|---|---|
|  |  |

Page intentionally left blank

Lesson 9

Color the fish and the coral reef. Cut out the fish. Tape or paste the fish into the coral reef.

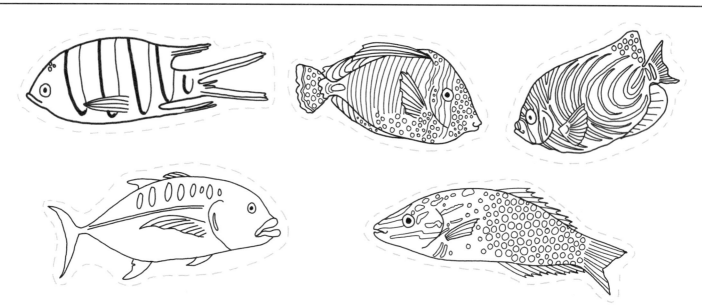

© GOOD AND BEAUTIFUL

Color the squares with a 1 blue. Color the squares with a 2 green. Parent tip: Color the images of the crayons with the correct colors for the child to refer to.

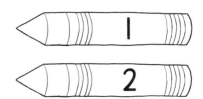

# Lesson 10

Draw a line connecting each mother shark to its pup. Color the sharks.

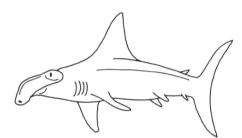

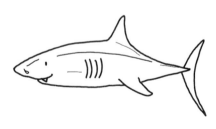

Draw lines to connect the matching sharks. Color the sharks.

Draw a line from each fish to the number of spots it has. Color the fish.

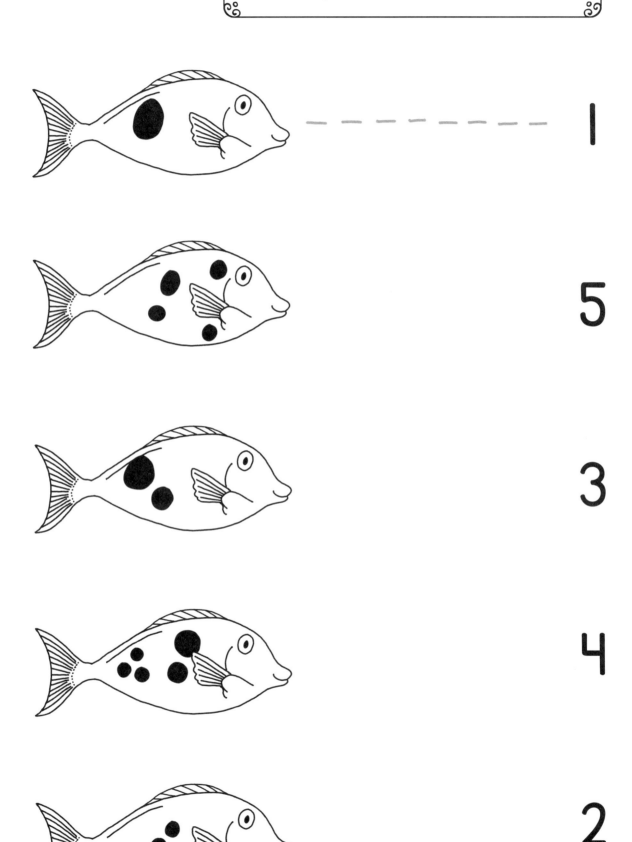

**Lesson 12**

Circle the six items in the top picture that you do not see in the bottom picture. Color the top picture.

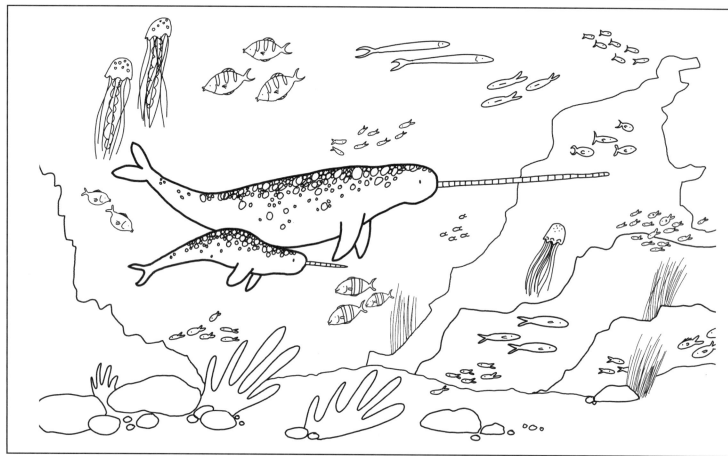

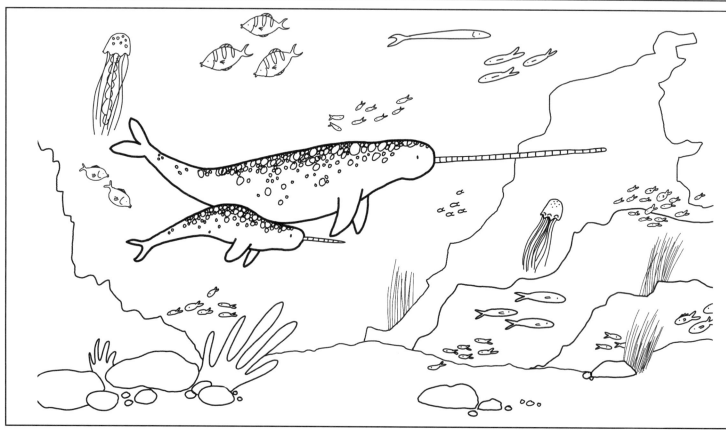

In each row, color the two creatures that are the same size.

# Extra Doodling and Drawing Page

# Extra Doodling and Drawing Page

# Extra Doodling and Drawing Page

# 1 Network
## GET CONNECTED

Tom Hutchinson
Kristin Sherman

with Mike Boyle

## ...ected

Social networking isn't ~~bad~~, but a change in the way we communicate. By providing opportunities for social networking and in-class sharing, **Network** is the first course to develop networking skills for language success in school and life, and at work.

## Social Media

**Your Network** activities build your English communication skills by inviting you to connect in class and online to share information about your new friends.

### YOUR NETWORK

 **IN CLASS:** Find someone in your class who can tell you about a new vacation place. Ask about three fun things you can do and see there.

**ONLINE:** Tell a partner about someone from your social network. What place did he/she recommend for you to visit? What are three things you can do and see there? You can share a picture of this place.

**Get Connected** lessons build your social media skills—from writing a personal profile to improving your 'netiquette' in English. Practice each skill until you're ready to take it online!

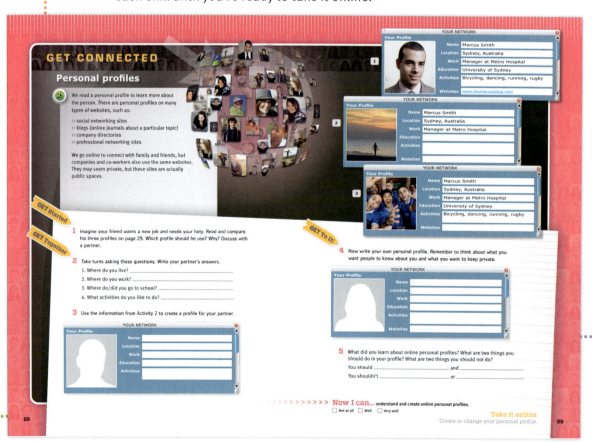

## Broadcast Video

**BBC video** and other authentic documentaries help you develop real-world listening skills, improve your English, and connect with the world around you!

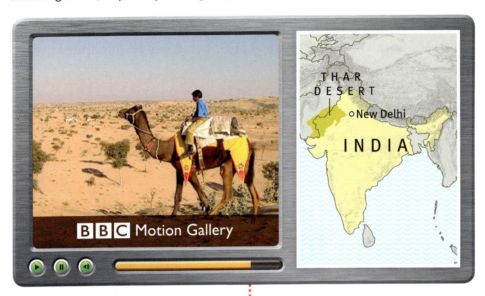

## Online Practice

**Network Online Practice** features over 100 engaging self-study activities to help you improve your vocabulary, grammar, speaking, and listening skills.

Use the **access card** on the inside back cover to log in at www.oxfordlearn.com/login.

Video supplied by BBC Motion Gallery

# Meet the *Network* characters

The *Network* storyline presents practical language you need for work, school, and life. The characters use networking, including social networking, to solve problems—just as you can in your life! They model real-world networking to show how we meet people—both in person and online.

In Student Book 1, you'll learn how Lucy uses social networking to find a new job, find out that Sarah meets someone special, and see what happens when Jordan's brother moves to town.

Tim

Ken

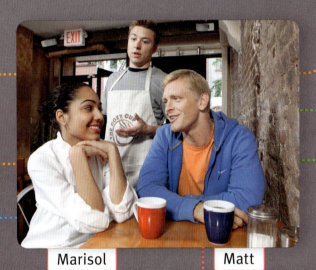
Marisol  Matt

Yuka

Jordan  Lucy

# Scope and Sequence

| | Unit | Conversation | Vocabulary | Grammar |
|---|---|---|---|---|
| 1 | **Where are you from?** Pages 2–7 | Talking about where people are from<br>Everyday expressions: *Saying hello and goodbye* | Countries and nationalities: *Brazil, Brazilian*, etc.<br>Language note: Countries and languages<br>Pronunciation: Word stress | *be*—statements and *yes/no* questions |
| 2 | **Everyday life** Pages 8–13 | Talking about daily routines<br>Everyday expressions: *Responding* | Daily routines: *get up, go to work*, etc.<br>Pronunciation: /ɪ/ and /i/ | Present simple—statements<br>Language note: Adverbs of frequency |
| 3 | **What do you like to do?** Pages 14–19 | Talking about what you like to do<br>Everyday expressions: *Talking on the phone* | Free-time activities: *play soccer, go to the movies*, etc.<br>Language note: *play* | Present simple—questions |
| 4 | **Meet the family.** Pages 20–25 | Asking about family<br>Everyday expressions: *Making suggestions* | Family members: *grandparents, niece*, etc.<br>Pronunciation: Unstressed vowel | *have/has*—statements, questions, and short answers<br>Language note: *some/any* |
| | **Review** Units 1–4 Pages 26–27 | • Vocabulary and Grammar self-assessment<br>• Reading: Edward from Vancouver | | |
| 5 | **Out on the town** Pages 30–35 | Asking about places in town<br>Everyday expressions: *Giving directions* | Places in a town: *a grocery store, a restaurant*, etc.<br>Language note: Articles (*a, the*) | *there is/there are*—statements, questions, and short answers |
| 6 | **Welcome home!** Pages 36–41 | Talking about what you're doing<br>Everyday expressions: *Offering and accepting help* | Rooms and furniture: *living room, sofa*, etc.<br>Language note: Prepositions of place<br>Pronunciation: The letter *a* | Present continuous—statements, questions, and short answers |
| 7 | **I can do that.** Pages 42–47 | Inviting someone to do something<br>Everyday expressions: *Talking about problems* | Months of the year: *March, August*, etc.<br>Language note: Months and dates | *can/can't*—statements, questions, and short answers |
| 8 | **I love my work.** Pages 48–53 | Explaining what you do<br>Everyday expressions: *Making polite requests* | Jobs: *mechanic, chef*, etc.<br>Pronunciation: Word stress | Present simple and present continuous |
| | **Review** Units 5–8 Pages 54–55 | • Vocabulary and Grammar self-assessment<br>• Reading: Home Sweet Home | | |

| Reading | ▶ Real-world Listening | ✳ Networking | Learning Outcomes Now I can... |
|---|---|---|---|
| Reading and speaking: *Welcome, new employees!* | Documentary: *The office* | Talk about a friend from another country | ...introduce someone and say where someone is from. ...name some countries and nationalities. ...say hello and goodbye. |
| Reading and writing: *A Day in My Life* Language note: *and, but, because* | **BBC** Motion Gallery *Life in the desert* | Ask about someone's daily routine | ...talk about my schedule. ...use the present tense and adverbs of frequency. ...respond to things people say. |
| Reading and writing: *Free-Time Survey* Language note: *like + -ing* | Documentary: *Yoga instructor* | Ask about someone's free-time activity | ...talk about my free time. ...ask questions with the present simple. ...understand and use phone expressions. |
| Reading and writing: *We're a Happy Family!* Language note: Object pronouns | Documentary: *Family barbecue* | Talk about brothers and sisters | ...talk about family members. ...use the verb *have/has*. ...make and respond to suggestions. |

**Get Connected ›››› Social Media Skill: Personal profiles**
Pages 28–29

| | | | |
|---|---|---|---|
| Reading and speaking: *Springfield Message Board* Language note: Pronouns for reference | **BBC** Motion Gallery *A different kind of town* | Ask someone about places in their town | ...talk about places around town. ...describe a town with *there is* and *there are*. ...ask for and give directions. |
| Reading and speaking: *A Room in a Day* | **BBC** Motion Gallery *Homes in Mongolia* | Talk about homes | ...talk about rooms and furniture. ...talk about what people are doing. ...offer and accept help. |
| Reading and writing: *You're Invited to a Barbecue!* Language note: *can* | Documentary: *Talent show audition* | Talk about holidays | ...talk about schedules and dates. ...use *can/can't* for ability, possibility, and permission. ...write an invitation. |
| Reading and writing: *On the Job* | **BBC** Motion Gallery *Mountain pilots* | Talk about jobs | ...talk about jobs. ...use the present simple and the present continuous. ...make and respond to polite requests. |

**Get Connected ›››› Social Media Skill: Getting started online**
Pages 56–57

Video supplied by BBC Motion Gallery

| Unit | Conversation | Vocabulary | Grammar |
|---|---|---|---|
| **9 Where were you?** Pages 58–63 | Talking about where people were<br>Everyday expressions: *Talking about bad news* | Events and places to go: *a concert, the movies,* etc.<br>Language note: *to/at* | Past simple: *to be*—statements, questions, and short answers<br>Pronunciation: *was/wasn't; were/weren't* |
| **10 What did you study?** Pages 64–69 | Talking about past classes<br>Everyday expressions: *Talking about good news* | School subjects: *physical education, history,* etc. | Past simple: statements—regular verbs, irregular verbs, and negative statements<br>Pronunciation: Past tense /ɪd/ endings |
| **11 What happened to you?** Pages 70–75 | Describing an injury<br>Everyday expressions: *Buying medicine* | Parts of the body: *neck, shoulder,* etc.<br>Pronunciation: Silent letters | Past simple: *yes/no* and *Wh-* questions |
| **12 I'm going on a cruise.** Pages 76–81 | Talking about vacation plans<br>Everyday expressions: *Showing surprise* | Travel: *luggage, passport,* etc.<br>Language note: *go* + preposition | *be + going to*—statements, questions, and short answers |
| **Review Units 9–12** Pages 82–83 | • Vocabulary and Grammar self-assessment<br>• Reading: Ann's Trip to England | | |
| **13 I eat a lot of cake.** Pages 86–91 | Talking about diet<br>Everyday expressions: *Talking at a restaurant* | Food: *cheese, tomatoes,* etc. | Count and noncount nouns<br>Pronunciation: *of* |
| **14 What do you like to wear?** Pages 92–97 | Talking about clothes<br>Everyday expressions: *Making comments* | Clothes: *a belt, a jacket,* etc.<br>Language note: plural words<br>Pronunciation: /s/ and /z/ | Adjectives |
| **15 My hometown is nicer.** Pages 98–103 | Talking about the weather<br>Everyday expressions: *Talking about a trip* | The weather: *sunny, cloudy,* etc.<br>Language note: Word building<br>Pronunciation: Vowel sounds | Comparatives: *colder, more dangerous,* etc. |
| **16 Around the world** Pages 104–109 | Talking about interesting places<br>Everyday expressions: *Asking for an explanation* | Geographical features: *a mountain, an island,* etc. | Superlatives: *the oldest, the most beautiful,* etc. |
| **Review Units 13–16** Pages 110–111 | • Vocabulary and Grammar self-assessment<br>• Reading: Jeff's Travel Blog | | |

| Reading | ▶ Real-world Listening | ✳ Networking | Learning Outcomes *Now I can...* |
|---|---|---|---|
| Reading and speaking: *Hotel Reviews* | BBC Motion Gallery *An American city* | Ask about interesting places to go | ...talk about places to go. ...say where people were. ...say how good or bad something was. |
| Reading and writing: *The New Eric* Language note: *could* | Documentary: *Hawthorne High* | Talk about school subjects | ...talk about my favorite school subjects. ...talk about the past. ...talk about good news. |
| Reading and speaking: *News from the Emergency Room* | BBC Motion Gallery *Laughter is the best medicine.* | Ask someone about an accident they had | ...name parts of the body. ...ask questions about past events. ...talk about buying medicine. |
| Reading and writing: *Big Plans!* Language note: Sequencers | BBC Motion Gallery *A trip to Honolulu* | Ask about travel ideas | ...talk about travel. ...describe plans for the future. ...show surprise. |

**Get Connected** ›››› **Social Media Skill:** Blogging
Pages 84–85

| | | | |
|---|---|---|---|
| Reading and speaking: *Breakfast around the World* Language note: Quantities | Documentary: *New York City food trucks* | Talk about breakfast food | ...talk about foods and drinks. ...talk about count and noncount nouns. ...order a meal in a restaurant. |
| Reading and speaking: *Work Clothes* | BBC Motion Gallery *Fashion and fabrics* | Ask someone about their favorite clothes to wear to parties | ...talk about clothes. ...describe things. ...make comments. |
| Reading and writing: *Seattle, My New Home* Language note: Compass directions | BBC Motion Gallery *Around the world—Chile* | Ask someone about the weather in their town | ...talk about the weather. ...compare two things. ...talk about preferences. |
| Reading and writing: *New Zealand, the most Beautiful Place in the World!* Language note: Paragraph planning | Documentary: *New York City landmarks* | Find someone who lives near an interesting place | ...talk about interesting places. ...talk about geographical features. ...describe a country. |

**Get Connected** ›››› **Social Media Skill:** Connecting online
Pages 112–113

Audio Scripts
pages 114–119

Grammar Reference
pages 120–127

Word List
pages 128–134

Irregular Verbs
page 135

Video supplied by BBC Motion Gallery

# UNIT 1 Where are you from?

## YOUR NETWORK

Go to Network Online Practice to record your voice in the conversations on pages 2 and 6.

Go to Network Online Practice to watch video about what different people do in an office.

Network! Go online to find someone from another country. Share on page 7.

Online Practice

### A CONVERSATION: I'm Brazilian.

1 Look at the picture. Where are the people?

2 Read and listen.
CD 1-02
- **Lucy:** Hi, Sarah. How are you?
- **Sarah:** Great. How are you?
- **Lucy:** Fine, thanks. I want you to meet my new co-worker, Tim.
- **Sarah:** Hi, Tim. Nice to meet you. I'm Sarah Ramiro.
- **Tim:** Hi, Sarah. It's nice to meet you, too. Where are you from?
- **Sarah:** Brazil. I'm Brazilian.

3 Work in a group of three. Practice the conversation.

4 Work in a new group of three. Practice the conversation again. This time use your real names.

> Hi, _____. How are you?

> Great. How are you?

> Fine, thanks. I want you to meet my new co-worker, _____.

**Now I can...** introduce someone and say where someone is from.
☐ Not at all  ☐ Well  ☐ Very well

## B VOCABULARY: Countries and nationalities

**1** Listen and repeat.

| | Country | Nationality | | Country | Nationality |
|---|---|---|---|---|---|
| | 1. Japan | Japanese | | 6. Argentina | Argentinian |
| | 2. Mexico | Mexican | | 7. Australia | Australian |
| | 3. Korea | Korean | | 8. Brazil | Brazilian |
| | 4. the United States | American | | 9. China | Chinese |
| | 5. Turkey | Turkish | | 10. Thailand | Thai |

Online Practice

**2** Write three more countries and nationalities.

| Country | Nationality |
|---|---|
| 1. | |
| 2. | |
| 3. | |

**Language note: Countries and languages**

1. Countries, nationalities, and languages always start with capital letters.
   I'm from **M**exico. I'm **M**exican. I speak **S**panish.
2. In most Latin American countries people speak Spanish. In Brazil they speak Portuguese.

    Spain    Spanish

    Portugal    Portuguese

**3** Listen to the conversations. Write the nationalities.

1. Hiroko is _Japanese_. Pedro is _____.
2. Ava is _____. José is _____.
3. Javier is _____. Laura is _____.
4. Sunan is _____. Andrew is _____.

### Pronunciation: Word stress

**Listen. Underline the stressed syllable.**

1. Ja•pan        Japa•nese        4. Korea        Korean
2. America       American         5. Canada       Canadian
3. Mexico        Mexican          6. Brazil       Brazilian

> > > > **Now I can...** name some countries and nationalities.
☐ Not at all  ☐ Well  ☐ Very well

UNIT 1 | Where are you from?

**GRAMMAR:** *be*—statements and *yes/no* questions

Grammar Reference page 120

| Affirmative statements | Negative statements | yes/no questions and short answers | |
|---|---|---|---|
| I'm Paul Smith. | I'm not from Mexico. | **Are** you Mexican? | Yes, I **am**. |
| You're late. | You **aren't** late. | | No, I'm not. |
| He's from New York. | She **isn't** from Boston. | | NOT ~~Yes, I'm.~~ |
| We're Brazilian. | We **aren't** Korean. | **Is** he Australian? | Yes, he **is**. |
| They're Japanese. | They **aren't** from Thailand. | | No, he **isn't**. |
| | | | NOT ~~Yes, he's.~~ |

 Online Practice

1  Look at the conversation on page 2. Find and circle the verb *be*.

 2  Listen to the affirmative statements. Write the negative statements.

**Example:**

You hear: *They're from New York.*

You write: *They aren't from New York.*

3  Complete the sentences with the correct verbs.

1. A: _____Are_____ you from Los Angeles?
   B: No, I _____. I _____ from Miami.
2. A: _____ she American?
   B: No, she _____. She _____ Canadian.
3. A: _____ they from Argentina?
   B: No, they _____. They _____ from Brazil.
4. A: _____ you Japanese?
   B: Yes, we _____. I _____ from Tokyo, and Yumiko _____ from Osaka.

Miami Beach

 4  Practice the conversations in Activity 3 with a partner.

5  Student A thinks of a famous person. Student B asks questions. Use the words below.

| a man/a woman | from | a movie star on TV | an athlete |
| young/old | married | a politician | a singer |

**Example:**

B: *Is the person a man?*

A: *No, she isn't.*

B: *Is she young?*

A: *Yes, she is.*

>  >  >  >  > **Now I can...** use the present tense with the verb *be*.
☐ Not at all  ☐ Well  ☐ Very well

## D READING AND SPEAKING

The blue words in the readings are *challenge words*. Go to Network Online Practice to practice them.

1 Read and listen.

### Intex International

### Welcome, new employees!

Please welcome Intex International's new employees around the world. Here are interviews with two of our new employees.

A: Hi. What's your name?
B: I'm Marisa Pereira.
A: Nice to meet you, Marisa. What's your new job?
5 B: I'm a translator in the Argentina office.
A: Oh, really? Are you bilingual?
B: Actually, I'm multilingual. I'm from Brazil, so I speak Portuguese. I also speak Spanish, English, and a little
10 Japanese.
A: Wow! Well, welcome to Intex International.
B: Thank you! I'm delighted to be here.

A: Hello. What is your name?
15 B: I'm David Johnson.
A: Welcome to Intex International, David.
B: Thanks! This is a wonderful company. Everyone is so friendly!
A: Yes, the people here are great. We're always very
20 busy, but we have a lot of fun. So, where are you from, David?
B: Well, I'm from Australia, but I work in the Korea office. My mother is from Korea, so I speak Korean.
A: Oh, I see. Are you the new sales manager?
25 B: Oh, no, I'm the new assistant. The sales manager is my boss.

Online Practice

2 Read again. Find the words in blue. Match each word with its meaning.

___ 1. bilingual     a. does many things
___ 2. multilingual     b. helper
___ 3. delighted     c. speaks two languages
___ 4. busy     d. very happy
___ 5. manager     e. boss
___ 6. assistant     f. speaks three or more languages

**Language note:** *Wh-* questions with *be*

*Wh-* questions have the same word order as *yes/no* questions.

| | |
|---|---|
| Is your name Teodor? | (*yes/no* question) |
| What is your name? | (*Wh-* question) |
| Are you 36? | (*yes/no* question) |
| How old are you? | (*Wh-* question) |

3 Interview a partner. Here are some questions you can ask:

| | | |
|---|---|---|
| What is your name? | What is your job? | Are you bilingual? |
| Where are you from? | Are you a student? | Who is your boss/teacher? |

> > > > **Now I can...** read and talk about personal information.
☐ Not at all    ☐ Well    ☐ Very well

## E YOUR STORY: Meet my friends.

**1** Read and listen to the story. Where are the people?

I'm Jordan Morris. I'm from Australia. I'm an actor. I'm also a waiter. Lucy is my girlfriend. This is her office. She's an assistant.

**Jordan:** Hi, Lucy. Are you busy?
**Lucy:** Hi, Jordan. Yes. I'm very busy!
**Olive:** Lucy? What's my schedule today?

Ryan Gaskell and his wife Cindy own this cafe, Cozy Cup. Ryan is American, and Cindy is Canadian. Their children, Melanie and Russell, aren't here now. Melanie is at school in England. Russell is on a trip to Mexico.

**Jordan:** Their names are Peter Colombo and Sarah Ramiro. Peter's from New York. He's a sales manager. Sarah is from Brazil. She's a student. They are both friends of Lucy.
**Sarah:** Hi, Jordan.
**Jordan:** Hi, Sarah. Hey, Peter, how's it going?
**Peter:** Hi, Jordan.

**Cindy:** Good morning, Peter. Coffee?
**Peter:** Yes, please.
**Cindy:** What about you, Sarah?
**Sarah:** No, thanks. I'm on my way to school.
**Cindy:** OK. See you later.

**2** Listen again. Match the first names with the descriptions.

  _b_  1. Jordan  a. American sales manager
  ___  2. Ryan    b. Australian actor
  ___  3. Sarah   c. American office assistant
  ___  4. Cindy   d. Canadian cafe owner
  ___  5. Lucy    e. Brazilian student
  ___  6. Peter   f. American cafe owner

**3** Listen to the expressions and repeat.

**4** Work in a group. Practice the story.

**Everyday expressions—Saying hello and goodbye**

| Good morning. | Goodbye. |
| Good afternoon. | Bye. |
| Good evening. | See you (later). |
| Hello./Hi. | Goodnight. |

> > > > > **Now I can...** say hello and goodbye.
☐ Not at all   ☐ Well   ☐ Very well

6

**REAL-WORLD LISTENING: The office**

1. Look at the photo. Where is the person? What is he doing?

2. Watch or listen to Mike talking about his office. Why does he like his job? Check ✓ the correct reason.

   ☐ a. He likes his job because it is a small company.
   ☐ b. He likes his job because the people are great.
   ☐ c. He likes his job because the office is in Miami.

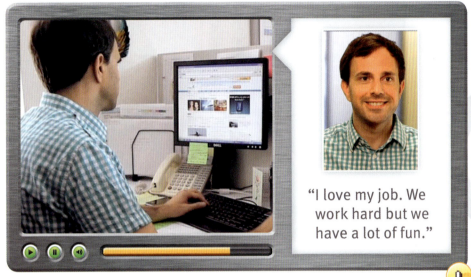

"I love my job. We work hard but we have a lot of fun."

Online Practice

3. Watch or listen again. Circle the correct words to complete the sentences.
   1. Mike works in *Canada / the main office*.
   2. Ben and *Josh / Kelly* are from Canada.
   3. Kate is the *office manager / office assistant*.
   4. Nancy is *American / from China*.
   5. Yasuko speaks both English and *Chinese / Japanese*.
   6. Chris is the *office assistant / sales assistant*.

4. Tell a partner about three people in your class or workplace.

   *Maria is our office assistant. She's from Miami and she speaks both English and Spanish.*

**YOUR NETWORK**

**IN CLASS:** Introduce yourself to your partner. Tell him/her your first and last name and your nationality.

**ONLINE:** Tell a partner about someone from your social network. What country is the person from? What is his/her name? You can share a picture of this person.

> > > > **Now I can...** give information about the people that I work or study with.
   ☐ Not at all   ☐ Well   ☐ Very well

# UNIT 2 Everyday life

## YOUR NETWORK

Go to Network Online Practice to record your voice in the conversations on pages 8 and 12.

Go to Network Online Practice to watch video about life in the desert.

Network! Go online to learn about someone's daily routine. Share on page 13.

Online Practice

### A CONVERSATION: Are you free today?

1 Look at the picture. What are Yuka and Sarah doing?

2 Read and listen.
CD 1-12

**Yuka:** Let's study for the test together.
**Sarah:** OK. When?
**Yuka:** Is tomorrow at 9:00 a.m. OK?
**Sarah:** No, my class is every morning from 9:00 to 10:00.
**Yuka:** Hmm. Is tomorrow at 1:00 p.m. good?
**Sarah:** I eat lunch at 1:00 with Lucy on Tuesdays. Are you free at 2:30 today?
**Yuka:** Yes, I think so. I'm not so busy!

3 Work in pairs. Practice the conversation.

4 Work in pairs. Practice the conversation again. This time, replace the times with new times.

> Let's study for the test together.
>
> OK. When?

> Is tomorrow at _____ OK?
>
> No, my class is every morning from _____ to _____.

**Now I can...** talk about my schedule.
 Not at all   Well   Very well

**VOCABULARY:** Daily routines

 1  Listen and repeat. Then check ✓ the activities you do every day.

☐ 1. get up

☐ 2. go to work

☐ 3. eat lunch

☐ 4. finish work

☐ 5. watch TV

☐ 6. go to bed

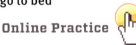

2  Think of more daily activities. Share them with a partner.

3  Write sentences about your daily routine.

*I get up at 7:30.*

1. I _____ at _____.    3. I _____ at _____.
2. I _____ at _____.    4. I _____ at _____.

4  Work with a partner. Talk about your daily routine.

I get up at 7:30. What about you?

I get up at 7:00.

### Pronunciation: /ɪ/ or /iː/

 1. Listen and repeat.

| /ɪ/ | /i/ |
|---|---|
| 1. live | leave |
| 2. his | he's |
| 3. it | eat |
| 4. this | these |

 2. Listen. Circle the word you hear in each pair above.

 > > > > > **Now I can...** talk about my daily routine.
☐ Not at all   ☐ Well   ☐ Very well

UNIT 2 | Everyday life

## C GRAMMAR: Present simple—statements

CD 1-16

**Grammar Reference page 120**

| Affirmative statements | | Negative statements |
|---|---|---|
| I **work** in a bank. | She **works** in a bank. | I **don't like** my job. |
| You **cook** dinner. | He **cooks** dinner. | He **doesn't like** his job. |
| They **watch** TV. | He **watches** TV. | NOT ~~He doesn't likes his job.~~ |

Online Practice

1. Look at the conversation on page 8. Find and circle the verbs in the present tense.

2. Complete the sentences with the correct affirmative form of the verb in parentheses.
   1. I _____ (go) to work at 8:00.
   2. She _____ (get up) at 7:00 every morning.
   3. They usually _____ (eat) a big lunch.
   4. I _____ (work) at a restaurant.
   5. We _____ (watch) TV for an hour every night.
   6. He _____ (finish) work at 6:00.

 3. Listen to the affirmative statements. Write the negative statements.

   CD 1-17

   **Example:**
   You hear:
   *She works in a restaurant.*
   You write:
   *She doesn't work in a restaurant.*

   **Language note: Adverbs of frequency**
   I **always** get up at 6:00.  0% ▓▓▓▓▓▓▓▓▓▓ 100%
   I **usually** go to work at 8:00.  0% ▓▓▓▓▓▓▓▓ 100%
   I **sometimes** go to a restaurant for lunch.  0% ▓▓▓ 100%
   I **never** eat lunch at home.  0% 100%

| Monday | Tuesday | Wednesday | Thursday | Friday |
|---|---|---|---|---|
| Get up!  7:30 | Get up!  7:30 | Get up!  6:30 | Get up!  7:30 | Get up!  7:30 |
| Work  9:00-5:00 | Work  9:00-5:00 | Work  9:00-5:00 | Work  9:00-7:00 | Work  9:00-5:00 |
| Dinner with Sam 7:00 | Watch TV  8:00-10:00 | Watch TV  8:00-10:00 | Dinner with Cara 7:00 | Movies  9:00-11:00 |
| Go to bed  11:00 | Go to bed  10:00 | Go to bed  10:00 | Go to bed  11:00 | Go to bed  12:00 |

4. Look at John's schedule for a typical week. Complete the sentences with a verb and an adverb of frequency.
   1. John _____usually_____ _____gets up_____ at 7:30.
   2. He _____ _____ to work at 9:00.
   3. He _____ _____ work at 5:00.
   4. He _____ _____ dinner with a friend.
   5. He _____ _____ TV in the evening.
   6. He _____ _____ to bed at 9:30.

> > > > > **Now I can...** use the present tense and adverbs of frequency. > > >
☐ Not at all  ☐ Well  ☐ Very well

# D  READING AND WRITING

1  Read and listen. What is Ligaya's job?

## A Day in My Life

My name is Ligaya Ragas, and I **live** in Davao, in the Philippines. I'm a teacher. I **teach** English in an elementary school. I don't live near the school. It's downtown, and houses there are expensive.
5 I take a jeepney (a kind of bus) to work. It takes a long time because the traffic is always very bad, and in the summer it's very hot.

School starts at seven-thirty in the morning. I get up at quarter to five, and I usually leave
10 home at 6:00. I **walk** to the main road and **wait** for the jeepney. I finish work at 5:00, and then I usually go home. I sometimes go grocery shopping first. My husband usually cooks dinner, and I **wash** the dishes.

15 I **like** my job, but I don't **earn** a lot of money, and I don't like the long trip to work every day.

I don't usually work on weekends. On Saturdays, I usually go shopping. On Sundays,
20 I often go to the beach with my husband and children. On rainy Sundays, we sometimes **stay** home and watch
25 movies.

**Online Practice**

2  What is the reading about? Check ✓ the best sentence.
☐ 1. The reading is about what Ligaya does every day at work.
☐ 2. The reading is about how Ligaya goes to work.
☐ 3. The reading is about why Ligaya likes her job.
☐ 4. The reading is about what Ligaya does every day.

**Language note:** *and*, *but*, *because*

My name's Ligaya, **and** I live in Davao.

I like my job, **but** I don't earn a lot of money.

It takes a long time **because** the traffic is bad.

3  Write about your life.

My name's _____ and I live in

_____. I _____.
(job/study)

I _____ to work/school every day. I start work/school at _____.
(drive/take the ___/walk)

I usually get up at _____ and I go to work at _____. I finish work

at _____ and then I _____. I like/don't like my job/school because

_____. On Saturday, I sometimes _____. On Sunday, I usually

_____.

> > > > > **Now I can...** write about my life.
☐ Not at all  ☐ Well  ☐ Very well

UNIT 2 | Everyday life   11

## E YOUR STORY: Sarah likes to exercise.

1 Read and listen to the story. Why does Peter come to the cafe?

**Ryan:** Hello, Peter. How are you?
**Peter:** I'm fine, thanks. And you?
**Ryan:** Fine.
**Peter:** Is Sarah here?
**Ryan:** No, she isn't. What time is it?
**Peter:** It's 6:30.
**Ryan:** She usually comes in at this time.

**Cindy:** Sarah? She swims on Tuesday evenings now.
**Ryan:** Swims? Wow, she really likes to exercise.
**Peter:** Yes, I know. She runs every day, too.
**Ryan:** Anyway, would you like a cup of coffee, Peter?
**Peter:** Oh, uh… no, thanks. I can't stay. See you.

**Ryan:** That's funny. Peter comes into the cafe, but he doesn't want coffee.
**Cindy:** He wants to see Sarah, but she isn't here.
**Ryan:** Oh, right.
**Cindy:** I think he likes her.
**Ryan:** Of course. They're friends.
**Cindy:** No, I think he *really* likes her.
**Ryan:** Really?

Ryan! You never notice anything.

2 Read the statements. Listen again. Write true (**T**) or false (**F**).

____ 1. It's 6:00 in the evening.
____ 2. It's Tuesday.
____ 3. Sarah is at the cafe.
____ 4. Sarah swims every day.
____ 5. Peter doesn't have a cup of coffee.
____ 6. Cindy thinks that Peter likes Sarah.

3 Use the story to complete the expressions in the box. Listen and check.

4 Work in a group. Practice the story.

**Everyday expressions—Responding**

Yes, I _____.
That's _____.
Oh, _____.

>>>> **Now I can...** respond to things people say. >>>>

☐ Not at all  ☐ Well  ☐ Very well

# REAL-WORLD LISTENING: Life in the desert

1. Look at the photo and map. What do you know about this place? Are there places like this near you?

2. What do you think life is like in the Thar Desert? Check ✓ the things you think people usually do there.

   ☐ go grocery shopping     ☐ watch TV
   ☐ have English class      ☐ keep goats and cows
   ☐ take a camel to school  ☐ walk to get water

3. Watch or listen and check your answers.

4. Watch or listen again. Who does these things? Check ✓ the correct answer.

   |              | Indojit | His father | His mother | Everyone |
   |--------------|---------|------------|------------|----------|
   | 1. milk goats | ☐ | ☐ | ☐ | ☐ |
   | 2. get water  | ☐ | ☐ | ☐ | ☐ |
   | 3. go to school | ☐ | ☐ | ☐ | ☐ |
   | 4. cook dinner | ☐ | ☐ | ☐ | ☐ |
   | 5. sit and talk | ☐ | ☐ | ☐ | ☐ |
   | 6. make shoes | ☐ | ☐ | ☐ | ☐ |

5. Look at the things in Activity 4. How often do you do these things? Tell a partner.

**YOUR NETWORK**

**IN CLASS:** Tell a partner about what you do on weekdays. What do you do in the morning, in the afternoon, and in the evening?

**ONLINE:** Tell a partner about someone from your social network. What is his/her daily routine? Share three interesting things about his/her routine. You can share a picture of this person.

> > > > > **Now I can...** talk about someone else's routine.
☐ Not at all  ☐ Well  ☐ Very well

# UNIT 3 What do you like to do?

## YOUR NETWORK

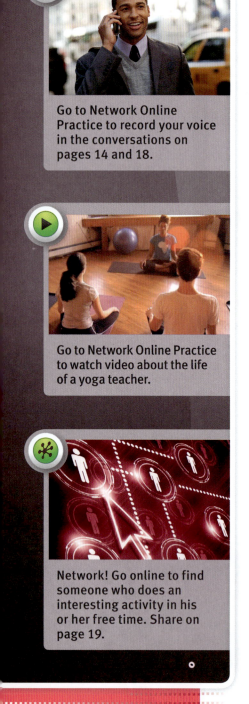

Go to Network Online Practice to record your voice in the conversations on pages 14 and 18.

Go to Network Online Practice to watch video about the life of a yoga teacher.

Network! Go online to find someone who does an interesting activity in his or her free time. Share on page 19.

Online Practice

### A CONVERSATION: Do you like sports?

1. Look at the picture. What is Sarah doing?

2. Read and listen.
   CD 1-22
   **Yuka:** Do you do yoga every day, Sarah?
   **Sarah:** No, I don't. I practice two or three times a week.
   **Yuka:** Do you go to the gym, too?
   **Sarah:** Yes, I do. I like a lot of sports. Do you like sports?
   **Yuka:** Sure. I like to watch them on television!

3. Work in pairs. Practice the conversation.

4. Work in pairs. Ask and answer the questions with a partner. Give information about yourself.

   Do you do yoga every day?
   Yes, I do./No, I don't.

   Do you like sports?
   Yes, I like _____./No, I don't.

   Do you go to the gym?
   Yes, I do./No, I don't.

**Now I can...** talk about my free time.
☐ Not at all  ☐ Well  ☐ Very well

## B  VOCABULARY: Free-time activities

**1** Listen and repeat. Then check ✓ the activities you like to do.

CD 1-23

☐ 1. visit with friends

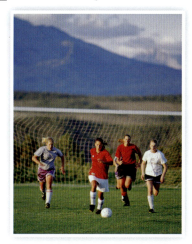
☐ 2. play soccer

☐ 3. play the piano

☐ 4. go to the movies

☐ 5. go to the gym

☐ 6. do yoga

**2** Write more free-time activities in the chart.

| Sports | Music | Other |
|---|---|---|
| play tennis | play the violin | do karate |
|  |  |  |
|  |  |  |

**Online Practice**

**Language note:** *play*

I **play** soccer. (*a sport*)
I **play** the piano.
(*a musical instrument*)

**3** Listen. What free-time activities does each person do? Write each activity under the correct name.

CD 1-24

1. Maria
_____
_____
_____

2. Joshua
_____
_____
_____

3. Anna
_____
_____
_____

**4** Share your answers to Activity 3 with a partner.

**Example:** *Maria swims. She also* _____ *and* _____.

> > > > > **Now I can...** talk about free-time activities.
☐ Not at all   ☐ Well   ☐ Very well

UNIT 3  |  What do you like to do?   15

## GRAMMAR: Present simple—questions

**Grammar Reference page 121**

| yes/no questions and short answers | | Wh- questions |
|---|---|---|
| Do you go to the gym? | Yes, I do. | What do you do in your free time? |
| Do they like sports? | No, they don't. | Where does he play soccer? |
| Does he play the piano? | Yes, he does. | |
| Does she run? | No, she doesn't. | |
| NOT ~~Does she runs?~~ | | |

Online Practice

1  Look at the conversation on page 14. Find and circle questions with the present simple.

2  Complete the conversation. Use the words in the box.

| do | does | don't | doesn't | run |
|---|---|---|---|---|

**Jack:** ___Do___₁ you ___run___₂ every day, Amelia?

**Amelia:** Yes, I _____₃.

**Jack:** _____₄ you _____₅ by yourself?

**Amelia:** No, I _____₆. I _____₇ with my friends.

**Jack:** _____₈ Abby _____₉ with you?

**Amelia:** No, she _____₁₀. I _____₁₁ with some friends from college.

**Jack:** Oh, _____₁₂ they live near you?

**Amelia:** No, they _____₁₃. We _____₁₄ in the park, so we meet there.

 3  Practice the conversation in Activity 2 with a partner.

4  Write questions with the phrases below.
   1. What/do/in your free time?  _What do you do in your free time?_
   2. like sports? _____
   3. play a musical instrument? _____
   4. Where/meet your friends? _____
   5. go to the gym every week? _____
   6. play computer games? _____

 5  Interview a partner. Use the questions from Activity 4.

6  Find a new partner. Ask about your first partners.

>>>> **Now I can...** ask questions with the present simple.
☐ Not at all  ☐ Well  ☐ Very well

## D  READING AND WRITING

1. Read and listen. Answer the questions in the survey.

### Free-Time Survey

Most American adults (96%) do a **leisure** activity (such as watching TV, visiting with friends, or exercising) every day. Are you similar to a **typical** American? Take this survey to find out!

1. How many hours of free time do you have every day?
   - a. 0
   - b. 1-2
   - c. 3-4
   - d. 5 or more

If your answer is d, you are like a typical American. The **average** American has 5.25 hours of leisure time every day. People age 75 and older have the most free time (7.8 hours), and people age 35-44 have the least free time (4.3 hours). Men have more free time (5.8 hours) than women do (5.1 hours).

2. What leisure activity do you do the most?
   - a. **exercise**
   - b. watch TV
   - c. **play computer games**
   - d. read
   - e. visit with friends
   - f. other _____

For the average American, the answer is b. Americans watch TV an average of 2.8 hours a day. Visiting with friends is the second most common leisure activity. People **socialize** for about 45 minutes per day.

3. Do you like exercising? How many hours a day do you exercise or do sports?
   - a. 0
   - b. 1-2
   - c. 3-4
   - d. 5 or more

On an average day, only 21% of men and 16% of women exercise. Most people are probably watching TV!

**Online Practice**

2. Read the statements. Write true (**T**) or false (**F**).
   - ____ 1. About half of Americans do a leisure activity every day.
   - ____ 2. The average American has 5.25 hours of free time every week.
   - ____ 3. Men have more free time than women.
   - ____ 4. Watching TV is the most common leisure activity for Americans.
   - ____ 5. Most people exercise every day.
   - ____ 6. I have more free time than the average American.

3. Write about yourself.

   I have _____ hours of free time every day. In my leisure time, I like _____, _____, and _____. I don't like _____ or _____. Every day, I watch _____ hours of TV and use the computer for _____ hours. I exercise _____ hours every week. I _____ a lot of free time!

**Language note:** like + -ing

I **like** watch**ing** TV.
I don't **like** read**ing**.
NOT I like watch TV.
OR I don't like read.

4. Share your paragraph with a partner.

### > > > > Now I can... write about my free-time activities.
☐ Not at all   ☐ Well   ☐ Very well

UNIT 3 | What do you like to do?

## E YOUR STORY: Meeting the new client

1 Read and listen to the story. What is the problem?

**Lucy:** Good afternoon. Martin, Hopper, and Green.
**Tim:** Hi, Lucy. This is Tim Johnson. Is Olive there?
**Lucy:** I'm sorry, she's out. Can I take a message?
**Tim:** Yes, please. I'm still in Toronto. The new client comes in at 2:00. Please ask Olive to call me back.
**Lucy:** OK. I'll tell her.

**Lucy:** Good afternoon. How can I help you?
**Olive:** Lucy, I'm out for the afternoon. I hurt my knee.
**Lucy:** But, Tim called... Hello? Olive? Hello?

**Tim:** Hello?
**Lucy:** Oh, Tim, thank goodness. Olive is out all afternoon.
**Tim:** That's not good. The client arrives there in five minutes.
**Lucy:** Oh, no!
**Tim:** Lucy, please take the new client out to eat. I'll call soon.
**Lucy:** OK, we can go to the Mexican restaurant.

2 Read the statements. Listen again. Write true (**T**), false (**F**), or no information (**N**).

_____ 1. Lucy is at work.
_____ 2. Olive is at work.
_____ 3. Tim likes Toronto.
_____ 4. Olive plays tennis.
_____ 5. Olive can't walk.
_____ 6. Tim asks Lucy to meet the client.

3 Use the story to complete some of the expressions in the box. Listen and check.

4 Work in a group. Practice the story.

**Everyday expressions—On the phone**

**Responses**
Who's calling, please?
I'm _____, she's out.
_____ I take a message?

**Reactions**
Yes, _____.
Please ask her to _____ me back.
Oh, well, can I leave a message, please?

Online Practice

>>>> **Now I can...** understand and use phone expressions.
☐ Not at all  ☐ Well  ☐ Very well

## F REAL-WORLD LISTENING: Yoga instructor

1. Look at the photos of Sarita. What do you think her job is? What does she do in her free time?

2. Watch or listen and complete the sentences about Sarita. Use the words in the box.

| every day | dancer | California | New York | yoga teacher |

Sarita Lou is from _____. She lives in _____. She's a _____. In her free time she is a _____. She does yoga _____.

"I love yoga because it's good for your health and it makes people happy."

Online Practice

3. Watch or listen again. Does Sarita do these things? Check ✓ Yes or No.

|  | Yes | No |
|---|---|---|
| 1. wake up at 8:00 | ☐ | ☐ |
| 2. do modern dance | ☐ | ☐ |
| 3. do ballet | ☐ | ☐ |
| 4. watch old TV shows | ☐ | ☐ |
| 5. have a lot of free time | ☐ | ☐ |
| 6. work in an office | ☐ | ☐ |

4. Do you exercise every day? What exercise do you do? Tell a partner.

### YOUR NETWORK

**IN CLASS:** Interview three classmates. What are three things each person likes to do in his/her free time?

**ONLINE:** Tell a partner about someone from your social network. What interesting thing does he/she like to do in his/her free time? You can share a picture of this person.

> > > > **Now I can...** talk about exercise.
☐ Not at all  ☐ Well  ☐ Very well

# UNIT 4 Meet the family.

## YOUR NETWORK

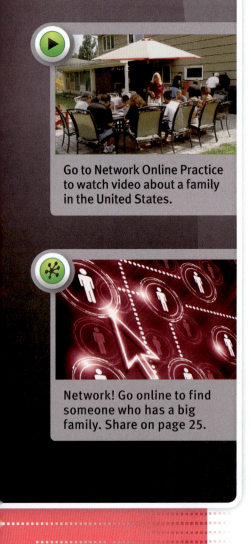

Go to Network Online Practice to record your voice in the conversations on pages 20 and 24.

Go to Network Online Practice to watch video about a family in the United States.

Network! Go online to find someone who has a big family. Share on page 25.

Online Practice

### A CONVERSATION: Do you have any brothers or sisters?

1 Look at the picture. What is Jordan looking at?

2 Read and listen.
CD 1-30

Lucy: Do you have any brothers or sisters, Jordan?

Jordan: Yes, I do. I have one sister and one brother. I have some photos in this box.

Lucy: Oh, your sister has very pretty blonde hair.

Jordan: No, she doesn't. She has brown hair…oh, that isn't my sister. That's Rachel, an old girlfriend. Now, why do I have that in there?

Lucy: Hmmm. I wonder.

3 Practice the conversation with a partner.

4 Work in pairs. Ask and answer questions about brothers and sisters. Try to ask your partner more questions.

Do you have any sisters/brothers?
Yes, I do./No, I don't.

How many sisters do you have?/Do you want a brother?

**Now I can...** talk about brothers and sisters.
☐ Not at all ☐ Well ☐ Very well

20

## B VOCABULARY: Family members

**1** Listen and repeat.

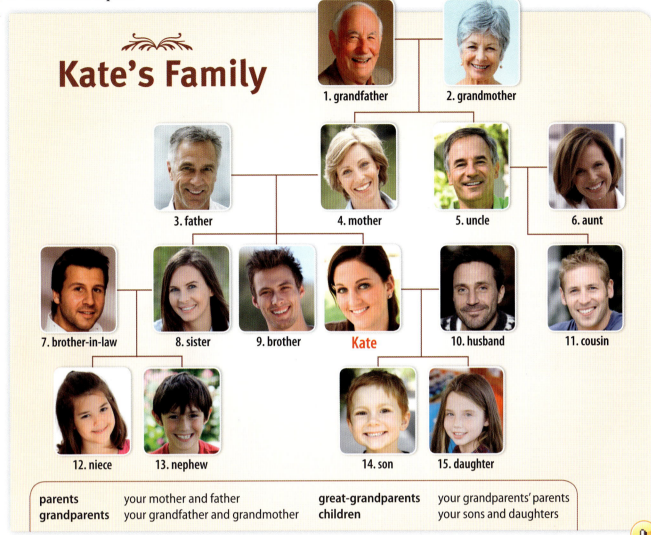

**2** Complete the sentences.

1. Your mother's sister is your ___aunt___.
2. Your father's parents are your _____.
3. Your uncle's children are your _____.
4. Your brother's son is your _____.
5. Your mother and father are your _____.
6. Your brother's daughter is your _____.

**3** Tell a partner about some people in your family.

**Example:** *My cousin's name is Ella. Her parents are my Uncle Luis and Aunt Maria. Ella is twenty-four. She's single.*

### Pronunciation: Unstressed vowel

A lot of English words have an unstressed vowel.

**1. Listen and repeat. Notice the unstressed vowel.**

br<u>o</u>ther    c<u>ou</u>sin    s<u>o</u>me    <u>u</u>ncle

**2. Listen. Circle the words with an unstressed vowel.**

mother    son    nephew    daughter
family    wife    niece    aunt

> > > > **Now I can...** talk about family members.
☐ Not at all  ☐ Well  ☐ Very well

UNIT 4 | Meet the family.

## C GRAMMAR: have/has
CD 1-34

| Statements | |
|---|---|
| We use *have/has* for: | |
| possessions | I **have** some photos. |
| families | We don't **have** any children. |
| descriptions | He doesn't **have** brown hair. |

**Grammar Reference page 121**

| Questions and short answers | |
|---|---|
| Do you **have** a pen? | Yes, I **do**. |
| | No, I **don't**. |
| Does he **have** blue eyes? | Yes, he **does**. |
| | No, he **doesn't**. |

Online Practice

| Questions with *How many* | |
|---|---|
| **How many** brothers do you have? | I have two brothers. |
| **How many** cousins does he have? | He has six cousins. |

1. Look at the conversation on page 20. Circle *have* and *has*.

2. Complete the sentences with *have* or *has*.
   1. How many sisters does he _____?
   2. We _____ five cousins.
   3. Do they _____ any brothers?
   4. She _____ short hair.
   5. I _____ a new car.
   6. Does he _____ a girlfriend?
   7. You _____ an older sister.
   8. Do you _____ any pets?

3. Write sentences about yourself. Use *have* or *don't have*.

   **Example:**
   *I have two brothers. I don't have any sisters.*

   1. (any) brother(s)
   2. (any) sister(s)
   3. a car
   4. (any) children
   5. long hair
   6. brown hair

 4. What do you have in your bag or backpack? Ask and answer questions with a partner.

   **Example:**
   A: *Do you have any pencils?*
   B: *Yes, I do. OR No, I don't.*
   A: *How many pencils do you have?*

   1. a pencil
   2. an English book
   3. any photos
   4. a cell phone
   5. a wallet or purse
   6. any keys
   7. a pen
   8. a drink

**Language note:** *some/any*

Use *some* with positive statements.
I have **some** photos.
Use *any* with negative statements and questions.
I don't have **any** photos.
Do you have **any** photos?

>>>>> **Now I can...** use the verb *have/has*.
☐ Not at all  ☐ Well  ☐ Very well

# D READING AND SPEAKING

**1** Read and listen. Who are the people in the photographs?

## We're a Happy Family!

My name's Roberto. I live with my wife, Mariluz, and our two children in Mexico City, but my **hometown** is Saltillo, a small town in the north of Mexico. My parents still live there.
5 They don't visit us very often because they don't like big cities. We visit them two or three times a year—at New Year's and in the summer—and when there's a big family **celebration** like a wedding or a special birthday.
10 I have a lot of **relatives** in Saltillo, too. My aunts, uncles, and cousins live there, so our visits are a lot of fun. My grandfather—that's my mother's father—is still alive. He's eighty-two, and he lives with one of my aunts.
15 My other grandparents are dead now.

I have a brother and a sister. My sister lives in Canada. I don't see her very often, but we **keep in touch** by e-mail and phone. She and her husband have one child—a little girl.
20 My brother is single. He lives in Mexico City, too, so I see him a lot. Our children love him. They think Uncle Manuel is cool because he drives a sports car.

Online Practice

**2** Read the text again. Answer the questions.
1. Where is Roberto's hometown?
2. Where does Roberto live now?
3. Do Roberto's parents like visiting him? Why or why not?
4. Does Roberto have any nieces or nephews? How many?
5. Is Roberto's brother married?

| **Language note:** Object pronouns |  |  |
|---|---|---|
| **Subject** | **Verb** | **Object** |
| We | visit | **them**. |
| They | don't visit | **us**. |
| She | calls | **him/her**. |
| He | doesn't use | **it**. |

**3** What does Roberto say about his family? Complete the sentences with the words in parentheses.
1. My brother's name is Manuel. Our children love ___him___ because _____ drives a sports car. (he / him)
2. _____ send e-mails to my sister and she often calls _____. (I / me)
3. My parents live in Saltillo. We see _____ two or three times a year, but _____ don't come to Mexico City. (they / them)
4. _____ see Manuel when he visits _____ on weekends. (we / us)
5. My sister is in Canada. I don't see _____ often. _____ has a daughter. (she / her)

**4** Tell a partner about your family.

*Where's your family from? How often do you see them? What do they like to do?*

> > > > **Now I can...** talk about my family.
☐ Not at all  ☐ Well  ☐ Very well

UNIT 4 | Meet the family.

**E** **YOUR STORY:** Sarah's family

1 Read and listen to the story. Who is in Sarah's family?

**Sarah:** Hi, Jordan. Do you have a date with Lucy?
**Jordan:** No, I don't. She does yoga on Mondays. It really helps her. She doesn't like her boss, Olive Green.
**Sarah:** Well, I don't have plans either. Why don't we get something to eat?
**Jordan:** Good idea. Let's get a sandwich here. Or how about soup?

**Sarah:** Mmm. This soup is very good. My mother makes wonderful soup.
**Jordan:** Do you come from a big family, Sarah?
**Sarah:** No, not really. I only have a brother and a sister.
**Jordan:** Are they both married?
**Sarah:** Yes, they are. And they both have children. So, I have nieces and nephews, too.

**Jordan:** Do you have a boyfriend back in São Paulo?
**Sarah:** Well, there's a guy there, Luis. He works in my family's business. My parents really like him.

**Peter:** Hello, Jordan. Hi, Sarah.
**Jordan:** Oh, hi, Peter. Why don't you join us?
**Peter:** Thanks, but I'm meeting some guys from work. See you later.
**Sarah:** Bye.

**Jordan:** I like Peter. He's a nice guy.
**Sarah:** Yes, he is nice. Very nice.

Online Practice

 2 Listen again. Complete the sentences.

1. _____ doesn't have a date with Lucy.
2. _____ doesn't like her boss.
3. _____ eats dinner with Jordan.
4. _____ makes wonderful soup.
5. _____ has a brother and a sister.
6. _____ are both married.
7. _____ is Sarah's friend in Brazil.
8. _____ is meeting some friends from work.

3 Use the story to complete the expressions in the box. Listen and check.

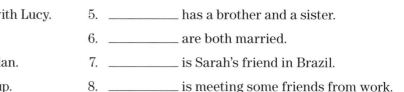

**Everyday expressions—Suggestions**

| Suggesting | Responding |
|---|---|
| Why _____ we get something to eat? | Good _____. |
| _____ get a sandwich here. | Thanks. |
| _____ about soup? | Yes, OK. |

 4 Work in a group. Do you agree with the sentences? Why or why not?

1. Family is important.
2. It is good to have a lot of brothers and sisters.
3. Family is more important than boyfriends or girlfriends.

> > > > > **Now I can...** make and respond to suggestions.
☐ Not at all  ☐ Well  ☐ Very well

# REAL-WORLD LISTENING: Family barbecue

1. Look at the photo of Donna and her family. Who are the people?

2. Watch or listen to Donna. Why is the family together today? Check ✓ the correct answer.
   - ☐ a. for a wedding
   - ☐ b. for a football game
   - ☐ c. for a birthday party

"We're going to have a good time today."

Online Practice

3. Watch or listen again. Match the person with the correct information.

   _____ 1. Lenny       a. is 12 years old today.
   _____ 2. Alexa       b. plays basketball, baseball, and football.
   _____ 3. Brandon     c. likes to skateboard.
   _____ 4. Jordan      d. is Donna's nephew.
   _____ 5. Lisa        e. is Donna's husband.
   _____ 6. Ryan        f. is Donna's sister.

4. Tell a partner about three people in your family. How often do you see them? When does your family get together?

## YOUR NETWORK

**IN CLASS:** Tell a partner about your favorite family member. How old is he/she? Where does he/she live? What does he/she do?

**ONLINE:** Tell a partner about someone from your social network. How many people are in his/her family? How many brothers/sisters does he/she have? Where is he/she from? You can share a picture of the family.

> > > > **Now I can...** give information about my family.
☐ Not at all   ☐ Well   ☐ Very well

UNIT 4 | Meet the family.

# REVIEW Units 1-4

Circle the correct word or words to complete each sentence.

## A | Vocabulary

1. Mayumi is from Tokyo. She is *Chinese / Japanese / Brazilian*.
2. Victor is American. He is from *the United States / Mexico / Turkey*.
3. I *eat / watch / go* a big lunch.
4. Lidia *finishes / goes / works* at a restaurant.
5. Walter *does / visits / plays* soccer.
6. We go to *the movies / soccer / friends* every Saturday.
7. Marta is my *uncle / niece / aunt*. She's my mother's sister.
8. My uncle has two sons. They are my *cousins / nephews / brothers*.

## B | Grammar

1. Elena and Sara are Mexican. They *isn't / am not / aren't* Brazilian.
2. Ming *am / is / are* Chinese.
3. A: Are you from Japan?
   B: Yes, I *is / are / am*.
4. They *don't work / doesn't work / isn't work* at home.
5. John gets up at 8:00 a.m. on Monday and Thursday. He *always / sometimes / never* gets up at 8:00 a.m.
6. Lisa *go / is go / goes* to work at 7:30.
7. *Do Ana go / Does Ana go / Go Ana* to the gym on Tuesdays?
8. A: Do they watch TV at night?
   B: Yes, they *do / does / watch*.
9. *What / What does / What do* you do in your free time?
10. Vera *have / has / haves* three brothers.
11. *How manys / How many / What many* sisters do you have?
12. A: Do you have any cousins?
    B: Yes, I *has / do / am*.

### Now I can...

- name some countries and nationalities (page 3)
- talk about my daily routine (page 9)
- talk about free-time activities (page 15)
- talk about family members (page 21)
- use the present tense with the verb *be* (page 4)
- use the present tense and adverbs of frequency (page 10)
- ask questions with the present simple (page 16)
- use the verb *have/has* (page 22)

## C | Reading

1  Read and listen to the story. What do Edward's family members like to do?

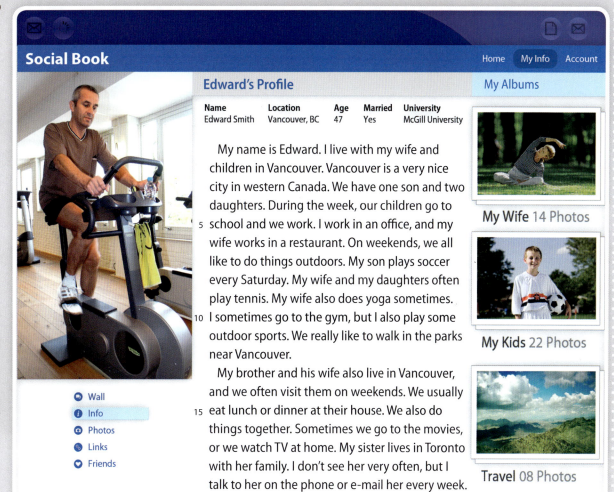

2  Read the story again. Answer the questions.

1. Where does Edward live?
2. How many people are in Edward's family?
3. Where does Edward work? Where does his wife work?
4. What does Edward like to do on weekends?
5. How often does Edward see his brother? How often does he see his sister?
6. Do you think Edward likes his life? Why or why not?

3  Who do these words refer to? Circle the correct answer.

1. I (line 1) _____
   A. Edward   B. Edward's brother   C. Edward's wife
2. we (line 6) _____
   A. Edward's brother and wife   B. Edward's family   C. Edward's son and daughters
3. his (line 13) _____
   A. Edward   B. Edward's son   C. Edward's brother
4. their (line 15) _____
   A. Edward's brother and wife   B. Edward's family   C. Edward's son and daughters
5. her (line 19) _____
   A. Edward's daughter   B. Edward's wife   C. Edward's sister

# GET CONNECTED

## Personal profiles

We read a personal profile to learn more about the person. There are personal profiles on many types of websites, such as:

›› social networking sites
›› blogs (online journals about a particular topic)
›› company directories
›› professional networking sites

We go online to connect with family and friends, but companies and co-workers also use the same websites. They may seem private, but these sites are actually public spaces.

**GET Started**

1. Imagine your friend wants a new job and needs your help. Read and compare his three profiles on page 29. Which profile should he use? Why? Discuss with a partner.

**GET Together**

2. Take turns asking these questions. Write your partner's answers.

   1. Where do you live? _____
   2. Where do you work? _____
   3. Where do/did you go to school? _____
   4. What activities do you like to do? _____

3. Use the information from Activity 2 to create a profile for your partner.

4. Now write your own personal profile. Remember to think about what you want people to know about you and what you want to keep private.

5. What did you learn about online personal profiles? What are two things you should do in your profile? What are two things you should not do?

You should _____ and _____.

You shouldn't _____ or _____.

> > > > > **Now I can...** understand and create online personal profiles.

☐ Not at all   ☐ Well   ☐ Very well

**Take it online**
Create or change your personal profile.

29

# UNIT 5 Out on the town

**YOUR NETWORK**

Go to Network Online Practice to record your voice in the conversations on pages 30 and 34.

Go to Network Online Practice to watch video about a different kind of town.

Network! Go online to learn about someone's neighborhood. Share on page 35.

Online Practice

## A  CONVERSATION: I need an ATM.

1 Look at the picture. Who do you see? What are they doing?

2 Read and listen.
CD 1-40

    **Yuka:** Excuse me. Are there any banks near here?
    **Woman:** Banks? No, there aren't. There are three or four banks downtown, but there isn't a bank near here.
    **Yuka:** Oh, no. I need an ATM.
    **Woman:** Oh, well, there's an ATM right there.
    **Yuka:** Oh, I see! Thank you very much.
    **Woman:** You're welcome.

3 Practice the conversation with a partner.

4 Work in pairs. Ask and answer questions about these places and your school or office.

| gyms | cafes | restaurants | banks |

Are there any _____ near here?
Yes, there are./No, there aren't.

Where are they?/Do you know where one is?

**Now I can...** ask about places in a town.
☐ Not at all  ☐ Well  ☐ Very well

30

## B  VOCABULARY: Places in a town

**1** Listen and repeat. Which places are in your neighborhood?

| | | | |
|---|---|---|---|
| 1. a park | 5. a restaurant | 9. a gas station | 13. a bank |
| 2. a grocery store | 6. a parking lot | 10. a drugstore | 14. a bus stop |
| 3. a gym | 7. an ATM | 11. public restrooms | 15. a cafe |
| 4. a subway station | 8. a square | 12. a hair salon | 16. a hotel |

Online Practice

**2** Think of more places in a town. Share them with a partner.

**3** Write the places.

1. You exercise here. ———
2. You buy medicine here. ———
3. You get money here. ———
4. You park your car here. ———
5. You buy food here. ———
6. You get a haircut here. ———

> **Language note:** Articles (*a*, *the*)
>
> Is **the** park near here?
> There is probably only one park in the town.
> Is there **a** park near here?
> There is probably more than one park in the town.

### Pronunciation: /s/ + consonants

**1. Listen and repeat.**

| station | square | restaurant |
| drugstore | bus stop | restroom |

**2. Say the sentences.**

1. The drugstore is in the mall.
2. Is there a restroom in this restaurant?
3. The bus stop is near the station.

> > > > **Now I can...** talk about places around town.
☐ Not at all   ☐ Well   ☐ Very well

UNIT 5 | Out on the town

## C GRAMMAR: *there is/there are*
CD 1-43

### Grammar Reference page 122

| Affirmative Statements | Negative Statements | Questions | |
|---|---|---|---|
| **There's** a drugstore on Main Street. | **There isn't** a drugstore near here. | **Is there** an ATM near here? | Yes, **there is.** |
| **There are** two banks in this town. | **There aren't** any banks near here. | | No, **there isn't.** |
| | **There aren't** any shops near here. | **Are there** any hotels in this town? | Yes, **there are.** |
| | | | No, **there aren't.** |

Online Practice

1 Look at the conversation on page 30. Circle statements and questions with *there is* and *there are*.

2 Complete the sentences about the town. Use *There is*, *There isn't*, *There are*, or *There aren't*.

1. _____ a hotel.
2. _____ two restaurants.
3. _____ a parking lot.
4. _____ two stores.
5. _____ any banks.
6. _____ a gym.
7. _____ any public restrooms.
8. _____ a gas station.

 3 Work with a partner. Ask and answer questions about the town in Activity 2.

**Example:**

A: *Is there a parking lot?*

B: *Yes, there is.*

A: *Are there any hotels?*

B: *No, there aren't.*

4 Work with a partner. Ask and answer questions about the area near your school. Use the words below.

**Example:**

A: *Is there an ATM near here?*

B: *Yes, there is. There's an ATM on Middle Street.*

- an ATM
- a gas station
- any public restrooms
- a gym
- any parking lots
- any good restaurants

>>>> **Now I can...** describe a town with *there is* and *there are*.  >>>
☐ Not at all  ☐ Well  ☐ Very well

# D READING AND SPEAKING

1  Read and listen. What does Ellen want to know?

2  Read the text again. Answer the questions.
   1. Why does Ellen want information about Springfield's neighborhoods?
   2. Do Lana, Calvin, and Mark like their neighborhoods? Why or why not?
      Write their reasons in the chart.

|  | Positive reasons ☺ | Negative reasons ☹ |
| --- | --- | --- |
| West End |  |  |
| Parkside |  |  |

3  Which neighborhood do you like better, the West End or Parkside? Why?

**Language note:** Pronouns for reference
There's a bank. **It's** on the corner.
There are some stores. **They're** big.

4  Talk to a partner. Describe your neighborhood. Use the words and expressions in the box.

| There's... | on the corner (of) | It's... | next to |
| There are... | across from | They're... | at the end of |

> > > > **Now I can...** describe my town or city.
   ☐ Not at all   ☐ Well   ☐ Very well

UNIT 5 | Out on the town

## E YOUR STORY: Lucy goes shoe shopping.

 1 Read and listen to the story. Where does Lucy want to go?

**1**

**Lucy:** Hi, Sarah, I don't have much time.
**Sarah:** It's your lunch hour, isn't it?
**Lucy:** Well, yes, but I'm running an errand for Olive.
**Sarah:** An errand?
**Lucy:** Yes, she needs shoes, and there's a great store down here.

**2**

**Lucy:** Excuse me. How do I get to Princess Shoes?
**Man:** Go down to the traffic light. Turn right. It's on the left.
**Lucy:** Thank you very much.

**3**

**Lucy:** Excuse me. Is Princess Shoes near here?
**Woman:** Princess Shoes? Hmmm. Go back that way, take the first left. Go past the bank. It's on the right, next to the post office.

**4**

**Lucy:** Hi, Sarah.
**Sarah:** Hi, Lucy. How's the shoe store?
**Lucy:** Great! I have Olive's new shoes and new shoes for me!
**Sarah:** Don't you have to get back to work?

Online Practice

 2 Listen again. Answer the questions.

1. What does Olive want?
2. How many people does Lucy talk to?
3. Who does Lucy buy shoes for?
4. What is next to the shoe store?

 3 Match the diagrams with the expressions. Listen and check.

 4 Work with a partner. Ask for directions to these places in your town.
- a park
- a bus station
- a drugstore
- a gym

**Everyday expressions—Giving directions**

___ Go past...
___ Turn right.
___ Turn left.
___ Take the first right.
___ Take the second left.
___ It's on the left.
___ It's on the right.
___ Go down...

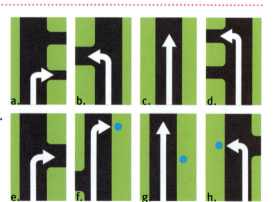

>>>> **Now I can...** ask for and give directions.
☐ Not at all  ☐ Well  ☐ Very well

#  REAL-WORLD LISTENING: A different kind of town

1. Look at the photo and map. What do you know about Florida? Why do people visit Florida?

2. Watch or listen to the report about a town in Florida called The Villages. What is different about this town?

Online Practice

3. Watch or listen again. What is there in The Villages? Check ✓ the items the town has.

- ☐ a square
- ☐ a TV station
- ☐ a lot of rain
- ☐ people under 18
- ☐ a swimming pool
- ☐ millions of retired people
- ☐ schools
- ☐ a tennis court
- ☐ a gym

4. Think about the older people in your town. What do they usually do? What places do they go to? Who do they usually live with? Tell a partner.

*In my country, older people often live with their children and grandchildren. They don't live in towns like The Villages. They don't usually go to the gym, but they sometimes...*

**YOUR NETWORK**

**IN CLASS:** Tell a partner about three interesting places in your neighborhood. What do you do there? Who do you go there with?

**ONLINE:** Tell a partner about someone from your social network. Where does he/she go with his/her friends in the neighborhood? You can share a picture of this place.

> > > > **Now I can...** talk about people in my town.
☐ Not at all  ☐ Well  ☐ Very well

UNIT 5 | Out on the town   35

# UNIT 6 Welcome home!

## YOUR NETWORK

Go to Network Online Practice to record your voice in the conversations on pages 36 and 40.

Go to Network Online Practice to watch video about homes in Mongolia.

Network! Go online to learn about someone's home. Share on page 41.

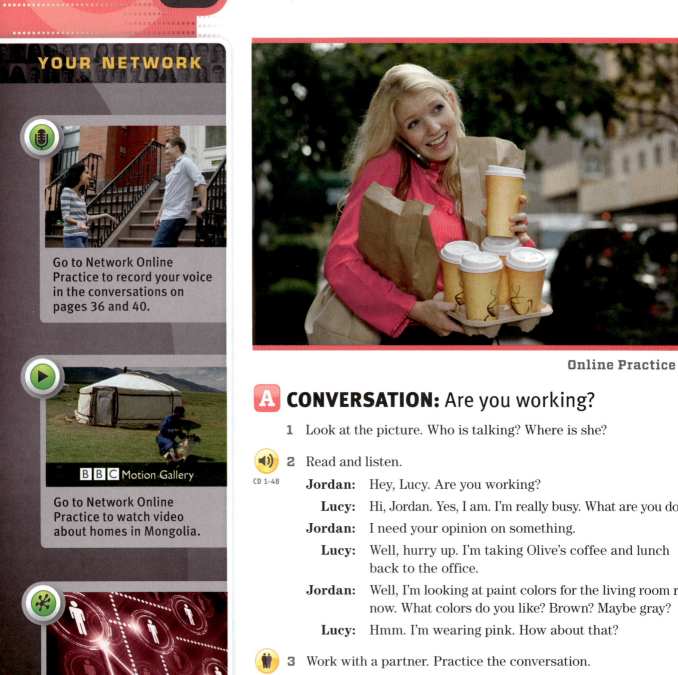

Online Practice

### A CONVERSATION: Are you working?

1. Look at the picture. Who is talking? Where is she?

2. Read and listen.
   CD 1-48

   **Jordan:** Hey, Lucy. Are you working?
   **Lucy:** Hi, Jordan. Yes, I am. I'm really busy. What are you doing?
   **Jordan:** I need your opinion on something.
   **Lucy:** Well, hurry up. I'm taking Olive's coffee and lunch back to the office.
   **Jordan:** Well, I'm looking at paint colors for the living room right now. What colors do you like? Brown? Maybe gray?
   **Lucy:** Hmm. I'm wearing pink. How about that?

3. Work with a partner. Practice the conversation.

4. Work in pairs. Ask and answer questions about what you are doing. Use the phrases in the box.

   | watching a movie | eating dinner |
   |---|---|
   | talking on the phone | drinking coffee |

   Are you working now?    What are you doing?
   No, I'm not.    I'm _____.

### Now I can... talk about what I am doing.
☐ Not at all  ☐ Well  ☐ Very well

36

## B  VOCABULARY: Rooms and furniture

 1  Listen and repeat.

1. living room
2. curtains
3. a sofa
4. an armchair
5. kitchen
6. a cabinet
7. a microwave
8. a refrigerator
9. a dishwasher
10. a stove
11. a table
12. bathroom
13. a bathtub
14. a mirror
15. a sink
16. bedroom
17. a rug
18. a dresser

2  What other furniture and rooms do you know? Can you name other things in the house?

1. _____
2. _____
3. _____
4. _____

**Language note:** Prepositions of place

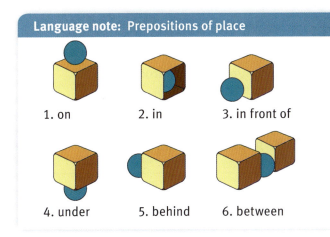

1. on   2. in   3. in front of

4. under   5. behind   6. between

 3  Tell a partner about a room in your home. Use prepositions from the box.

**Example:**

*In our living room, there are two armchairs in front of the TV. There's a table between the chairs.*

### Pronunciation: The letter *a*

1. In each line, circle the word with a different *a* sound.

| | | | |
|---|---|---|---|
| 1. paint | train | chair | dishwasher |
| 2. park | armchair | answer | party |
| 3. wash | watch | what | cash |
| 4. table | microwave | have | day |
| 5. water | later | make | take |
| 6. talk | last | walk | hall |

 2. Listen to the words in Part 1. Repeat. Were your answers correct?

> > > > > **Now I can...** talk about rooms and furniture.
☐ Not at all  ☐ Well  ☐ Very well

UNIT 6 | Welcome home!   37

## C GRAMMAR: Present continuous

| Affirmative statements | Negative statements | Questions and short answers | |
|---|---|---|---|
| I'm **cleaning** the kitchen. | I'm **not cleaning** the kitchen. | **Are** you **working**? | Yes, I **am**. |
| She's **watching** TV. | She **isn't watching** TV. | | No, I'm **not**. |
| They're **eating** lunch. | They **aren't eating** lunch. | **Is** he **painting** the bedroom? | Yes, he **is**. |
| | | | No, he **isn't**. |

Grammar Reference page 122

Online Practice

**1** Look at the conversation on page 36. Circle the verbs in the present continuous.

**2** Complete the sentences. Use the present continuous. Some of the sentences are negative.
1. We ___'re visiting___ with friends in the living room. (visit)
2. They _____ in the park. (not/run)
3. She _____. (work)
4. We _____ tennis. (not/play)
5. He _____ the bathroom. (not/paint)
6. I _____ English in my bedroom. (study)
7. They _____ dinner for their friends. (cook)
8. She _____ on the sofa. (not/sleep)

**3** Work with a partner. Use the phrases below to write questions and answers. Then practice the conversations.
1. you/paint/the bedroom? (Yes)
   A: _Are you painting the bedroom_?
   B: _Yes, I am_.
2. he/clean/the living room? (No)
   A: _____?
   B: _____.
3. she/make/coffee? (Yes)
   A: _____?
   B: _____.
4. they/playing soccer? (No)
   A: _____?
   B: _____.
5. you/study? (No)
   A: _____?
   B: _____.

**4** Listen to check your answers to Activity 3.

>>>> **Now I can...** talk about what people are doing. >>>
☐ Not at all  ☐ Well  ☐ Very well

## D READING AND SPEAKING

1 Read and listen. What kind of television show is Ben Wolf on?

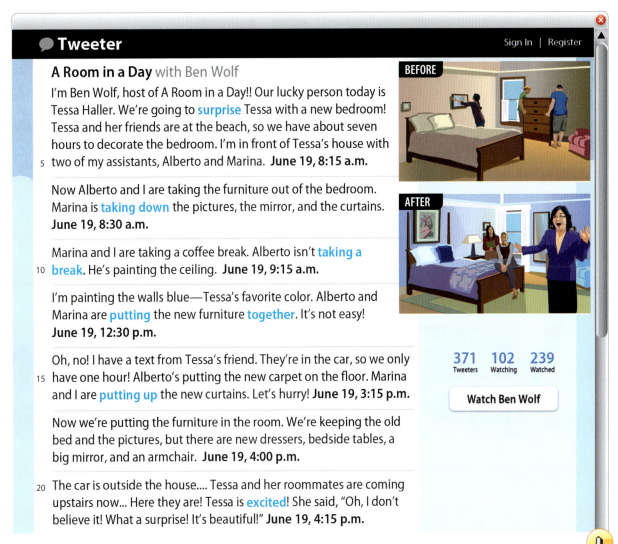

**Tweeter**　　　　　　　　　　　　　　　　Sign In | Register

**A Room in a Day** with Ben Wolf

I'm Ben Wolf, host of A Room in a Day!! Our lucky person today is Tessa Haller. We're going to surprise Tessa with a new bedroom! Tessa and her friends are at the beach, so we have about seven hours to decorate the bedroom. I'm in front of Tessa's house with
5　two of my assistants, Alberto and Marina. **June 19, 8:15 a.m.**

Now Alberto and I are taking the furniture out of the bedroom. Marina is taking down the pictures, the mirror, and the curtains. **June 19, 8:30 a.m.**

Marina and I are taking a coffee break. Alberto isn't taking a
10　break. He's painting the ceiling. **June 19, 9:15 a.m.**

I'm painting the walls blue—Tessa's favorite color. Alberto and Marina are putting the new furniture together. It's not easy! **June 19, 12:30 p.m.**

Oh, no! I have a text from Tessa's friend. They're in the car, so we only
15　have one hour! Alberto's putting the new carpet on the floor. Marina and I are putting up the new curtains. Let's hurry! **June 19, 3:15 p.m.**

Now we're putting the furniture in the room. We're keeping the old bed and the pictures, but there are new dressers, bedside tables, a big mirror, and an armchair. **June 19, 4:00 p.m.**

20　The car is outside the house.... Tessa and her roommates are coming upstairs now... Here they are! Tessa is excited! She said, "Oh, I don't believe it! What a surprise! It's beautiful!" **June 19, 4:15 p.m.**

**BEFORE**

**AFTER**

371 Tweeters　102 Watching　239 Watched

Watch Ben Wolf

Online Practice

2 Read the text again. Answer the questions.
   1. What is Ben Wolf's job?
   2. What is Ben doing today?
   3. What is Tessa doing today? Why?
   4. What furniture is new?
   5. How does Tess feel about her new bedroom?

 3 Talk to a partner. What is Ben doing at these times?

**Example:**
　A: *What is Ben doing at 9:15?*
　B: *He's taking a coffee break.*

| 9:15 | 12:30 | 4:00 | 8:30 | 3:15 | 4:15 |

4 Tell a partner about your ideal room.

In my ideal room, there's _____ (furniture).

The walls, the floor, and the curtains are _____ (color).

On the walls, there is/are _____ (pictures, a mirror).

> > > > **Now I can...** describe a room.
　☐ Not at all　☐ Well　☐ Very well

UNIT 6 | Welcome home!　39

## E YOUR STORY: Painting Jordan's apartment

1 Read and listen to the story. Why is everyone at Jordan's apartment?

**Sarah:** Hi, Jordan. Are you going to work?
**Jordan:** No, I'm going to the store for some paint.
**Sarah:** Are you painting your apartment?
**Jordan:** Yes, I am. I'm painting the living room.
**Sarah:** Would you like some help? I'm not doing anything today.
**Jordan:** That would be great. Thanks a lot. Lucy and Peter are coming to help, too.

**2 | LATER**

**Sarah:** Hello, Peter. Hi, Jordan. No Lucy?
**Jordan:** No, she's at home. She's writing her resume.
**Sarah:** Oh? Is she looking for a new job?
**Jordan:** Yes, she is. Anyway, we're moving the furniture right now.
**Sarah:** OK, let me give you a hand.
**Jordan:** Thanks... (phone rings) Excuse me.

**Jordan:** Hello? What? Now? But I'm doing something... OK.

**Jordan:** I'm sorry, but I have an audition now.
**Peter:** On a Sunday afternoon?
**Jordan:** Yeah, it's very last-minute.

**Sarah:** So, it's just you and me, Peter.
**Peter:** Yes, just you and me.

Online Practice

2 Listen again and answer the questions.

1. Why is Jordan going to the store?
2. Why are Sarah and Peter at Jordan's apartment?
3. Why isn't Lucy at Jordan's apartment?
4. Why does Jordan leave?
5. What day is it?

3 Use the story to complete the expressions in the box. Listen and check.

4 Work in a group. Retell the story.

**Everyday expressions—Offering and accepting help**

Would you _____ some help?
That would _____ great.
Thanks _____ lot.
Let me give you a _____.

> > > > > **Now I can...** offer and accept help.
☐ Not at all  ☐ Well  ☐ Very well

40

##  REAL-WORLD LISTENING: Homes in Mongolia

1. Look at the picture and the map. The home you see is a Mongolian *ger*. Do you think it is a nice home? What do you think are some good points and bad points about it?

2. Watch or listen to the report about homes in Mongolia. How long does it take to put up a Mongolian house? Why is this important?

Online Practice

3. Watch or listen again. How do animals help the people in Mongolia? Circle the correct answers.

    1. The people make a healthy drink from *horse's / sheep's* milk.
    2. The walls and roof come from *sheep / camel* wool.
    3. *Horses / Sheep* provide meat and wool.
    4. *Sheep / Horses* make fuel for the ovens.

4. How is a Mongolian *ger* different from your own home? Tell a classmate about three differences.

### YOUR NETWORK

**IN CLASS:** Tell a partner three things about your home. What rooms and furniture does it have? You can share a picture of your home.

**ONLINE:** Tell a partner about someone from your social network. What kind of home does he/she live in? Is it a house or an apartment? What rooms and furniture does it have? You can share a picture of his/her home.

> > > > > **Now I can...** talk about different kinds of homes.
☐ Not at all  ☐ Well  ☐ Very well

# UNIT 7
# I can do that.

## YOUR NETWORK

Go to Network Online Practice to record your voice in the conversations on pages 42 and 46.

Go to Network Online Practice to watch video about a man who can sing and play the guitar.

Network! Go online to learn about someone's favorite day of the year. Share on page 47.

Online Practice

## A  CONVERSATION: Time for an interview

1. Look at the picture. What is Lucy looking at?

2. Read and listen.
   CD 2-02

   **Assistant:** Can I speak to Lucy Patterson?
   **Lucy:** Yes, this is Lucy.
   **Assistant:** Hi, Lucy. I'm calling from Stevens Associates to schedule an interview.
   **Lucy:** Oh, that's great.
   **Assistant:** Can you come in on August 5?
   **Lucy:** Hmm. No, I'm sorry, I can't. I'm busy that day.
   **Assistant:** OK. Let's see… How about August 6?
   **Lucy:** Yes, I can do that. What time?

3. Work with a partner. Practice the conversation.

4. Work in pairs. Ask and answer questions about the activities in the box. Suggest a new date.

   | dinner party | soccer game | language class |

   Can you come to a _____ next Saturday?
   No, I can't.
   OK, how about next _____?
   Yes, I can do that./No, I can't.

**Now I can...** talk about schedules and dates.
☐ Not at all  ☐ Well  ☐ Very well

# B VOCABULARY: Months of the year

1. Listen and repeat.

| Months | | | | | |
|---|---|---|---|---|---|
| January | February | March | April | May | June |
| July | August | September | October | November | December |

## Calendar

**Ordinal Numbers**

| 1st | first |
| 2nd | second |
| 3rd | third |
| 4th | fourth |
| 5th | fifth |
| 6th | sixth |
| 7th | seventh |
| 8th | eighth |
| 9th | ninth |
| 10th | tenth |
| 11th | eleventh |
| 12th | twelfth |

**August**

| Sunday | Monday | Tuesday | Wednesday | Thursday | Friday | Saturday |
|---|---|---|---|---|---|---|
| 1 | 2 Dentist's appointment 10:00 | 3 | 4 | 5 | 6 | 7 Maria's party! 7:00 |
| 8 | 9 | 10 | 11 English Test! | 12 | 13 | 14 |
| 15 | 16 | 17 | 18 | 19 | 20 | 21 Soccer Game 4:00 |
| 22 | 23 | 24 | 25 | 26 Alison's birthday! | 27 | 28 |
| 29 | 30 | 31 | 1 | 2 | 3 | 4 |

2. Listen. Write the dates or months.
   1. The meeting is on ___April 16___.
   2. The English test is in _____.
   3. Kate's dentist's appointment is on _____.
   4. Helen is away until _____.
   5. Sam's wedding anniversary is on _____.
   6. John's birthday is in _____.

> **Language note: Months and dates**
> 1. September NOT ~~september~~
> 2. We write:
>    May 17 (or May 17th)
>    We say:
>    May seventeenth
> 3. We use:
>    **in** with **months**:
>    My birthday is **in** December.
>    **on** with **dates**:
>    My birthday is **on** December 2nd.

3. Work with a partner. Look at the calendar above. Ask and answer questions about Jack's activities.

   **Example:**
   A: *When is Jack's dentist's appointment?*
   B: *It's on August 2 at 10:00.*

4. Ask a partner these questions.
   1. What's the date today?
   2. What's the date next Tuesday?
   3. When's your birthday?
   4. What dates are important to you?

> > > > **Now I can...** talk about dates.
☐ Not at all  ☐ Well  ☐ Very well

## C GRAMMAR: can/can't

**Grammar Reference page 123**

| Statements | | |
|---|---|---|
| We use *can/can't* for: | | |
| ability | I **can** swim. | He **can't** swim. |
| possibility | He **can** play tennis today. | She **can't** play tennis today. |
| permission | You **can** park here. | You **can't** park here. |

| Questions and short answers | |
|---|---|
| **Can** you speak Japanese? | Yes, I **can**. |
|  | No, I **can't**. |
| **Can** he swim? | Yes, he **can**. |
|  | No, he **can't**. |

Online Practice

1. Look at the conversation on page 42. Circle the statements and questions with *can* and *can't*.

2. Complete the sentences about yourself. Use *can* or *can't*. Then write two more sentences with *can* or *can't*.

   **Example:**

   *I can read without glasses.* OR *I can't read without glasses.*

   1. I _____ read without glasses.
   2. I _____ run five miles.
   3. I _____ play tennis.
   4. I _____ swim a hundred yards.
   5. I _____ speak three languages.
   6. I _____ play a musical instrument.
   7. I _____.
   8. I _____.

3. Work with a partner. Ask and answer questions about the statements in Activity 2.

   **Example:**

   A: *Can you read without glasses?*
   B: *Yes, I can.* OR *No, I can't.*

4. Tell another student about your partner.

   **Example:**

   *Jack can run five miles. He can't play tennis.*

   ### Pronunciation: can/can't

   1. **Listen and repeat.**
      1. I can swim.   I can't swim.
      2. He can come on Monday.   He can't come on Monday.
      3. She can drive.   She can't drive.
      4. We can speak French.   We can't speak French.

   2. **Listen. Which sentence in each pair above do you hear? Circle it.**

>>>> **Now I can...** use *can/can't* for ability, possibility, and permission.   >>>>
☐ Not at all   ☐ Well   ☐ Very well

# D  READING AND WRITING

1. Read and listen. Why is Christine having a barbecue?

## You're Invited to a Barbecue!

Hi everyone –

I'm having a **barbecue** on May 30. There's no **special occasion** for the party... it's just a **celebration**. There will be something fun for everyone:

Do you like sports? There will be **volleyball** and soccer. Can you sing? We'll have karaoke. I know everyone can eat... and dance!

**Date:** Saturday, May 30
**Time:** 4:00-???
**Place:** My house

I hope you can make it! Please let me know by May 25. Thanks.

✓ Yes — **Amy:** That sounds like fun! Can I bring something? How about a dessert?

✗ No — **Ming:** Thanks for the invitation, Christine. Unfortunately, I can't **make it**. I'll be **out of town**. Can we have coffee sometime soon, though?

✓ Yes — **Jake:** Christine, the party sounds fun! I'm a terrible singer, so I can't do karaoke... but I can play volleyball and soccer! :-) See you on the 30th.

? Maybe — **Lila:** Hi, Christine—I'd love to come, but a friend is visiting that weekend. Can I bring her to the party?

✓ Yes — **Pedro:** Christine, I can't wait! Can you please give me your address?

Online Practice

2. Complete the following sentences.
   1. Amy is thinking about bringing _____ to _____.
   2. Ming can't go to the party because _____.
   3. Jake can't do karaoke because _____.
   4. Lila's friend is visiting, so she wants to _____.
   5. Pedro will go to the party, but he needs _____.

3. Write an invitation. Include this information:

   | What is it? | Where is it? |
   | Is it for a special occasion? | When is it? |

**Language note: can**

We also use *can* for offers and requests.
**Can** I bring a dessert?
**Can** you please give me your address?

> > > > > **Now I can...** write an invitation.
☐ Not at all  ☐ Well  ☐ Very well

UNIT 7  |  I can do that.    45

## E YOUR STORY: Jordan and Peter's problems

1 Read and listen to the story. Who are Peter and Jordan talking about?

**1**

**Jordan:** Aren't you going to your yoga class this evening?
**Lucy:** No, I'm not.
**Jordan:** But you go every Monday.
**Lucy:** Well, I'm not going this week. I'm worried about my interview.
**Jordan:** Well, yoga's good for stress.
**Lucy:** Look, I'm not going. All right?
**Jordan:** OK, OK. I'm only trying to help.

**2**

**Sarah:** I'm sorry, Peter. I really like you, but I can't go out with you.
**Peter:** I don't understand. What's the problem?
**Sarah:** I know it can't work.
**Peter:** Why not? Is it because I travel so much?
**Sarah:** No, it isn't. I'm sorry. I can't explain.

**3 | LATER THAT DAY**

**Jordan:** Lucy is worried about an interview for a new job. I'm only trying to help, but I always say the wrong thing.
**Peter:** Sorry, I can't help you there.
**Jordan:** What's your problem?
**Peter:** It's Sarah. She says she likes me, but she doesn't want to go out with me.
**Jordan:** Well, actually, Peter, I think I can help you there.

Online Practice

2 Read the statements. Listen again. Write true (**T**), false (**F**), or no information (**N**).

_____ 1. Lucy has an interview for a new job.
_____ 2. She isn't worried about the interview.
_____ 3. She is going to yoga next week.
_____ 4. Sarah doesn't like Peter.
_____ 5. Peter likes to travel a lot.
_____ 6. Jordan thinks he can help Peter.

3 Use the story to complete the expressions in the box. Listen and check.

4 Work with a partner. Talk about a problem you or your friend has.

**Example:**
*One problem I have is…*

**Everyday expressions—Problems**

I'm _____ about my interview.
I'm only trying to _____.
I _____ understand.
What's the _____?
I _____ explain.

> > > > > **Now I can…** talk about problems.
☐ Not at all   ☐ Well   ☐ Very well

# REAL-WORLD LISTENING: Talent show audition

1. Look at the photo of Emiliano. What can he do?

2. Watch or listen to Emiliano. Check ✓ the things he says he can do.

   ☐ play a musical instrument   ☐ speak two languages   ☐ play the blues
   ☐ write songs   ☐ dance   ☐ wait

Online Practice

3. Watch or listen again. Complete the words to Emiliano's song, "I Can't Have." Use the words in the box.

   | can't | dream | need | new | speak | want |
   |---|---|---|---|---|---|

   "You are the only girl I dream of,
   the only girl I _____ of,
   you're what I _____ have.
   I know that my mind should move on,
   groove on, find a _____ one,
   but I just can't shake you.

   You're what I can't have.
   I can't have all I ever _____.
   I can't have all I ever _____.
   All I want, all I _____,
   I'll never get the loving from you that I need."

4. Work in groups. Talk about a song you like. What are the words? Can you sing it? What's your favorite song in English? Why?

## YOUR NETWORK

**IN CLASS:** Tell a partner about your favorite day of the year. When is it? Why is it your favorite day? What do you do on that day?

**ONLINE:** Tell a partner about someone from your social network. What is his/her favorite day of the year? Why is it his/her favorite? What does he/she do on that day? You can share a picture of this person.

> > > > > **Now I can...** talk about a song I like.
   ☐ Not at all   ☐ Well   ☐ Very well

UNIT 7 | I can do that.

# UNIT 8 I love my work.

## YOUR NETWORK

Go to Network Online Practice to record your voice in the conversations on pages 48 and 52.

Go to Network Online Practice to watch video about mountain pilots.

Network! Go online to learn about someone's dream job. Share on page 53.

Online Practice

### A CONVERSATION: How do you like the job?

1 Look at the picture. Who do you think the new woman is?

2 Read and listen.
CD 2-12

**Cindy:** Marisol, this is Jordan. He's a server. He sometimes works here at lunchtime. Jordan, this is Marisol, our new chef. She started last week.

**Marisol:** Nice to meet you, Jordan.

**Jordan:** Nice to meet you, too. How do you like the job?

**Marisol:** I like it a lot. I usually prepare salads in the morning, but they didn't bring our vegetable order today. So I'm doing Sudoku.

**Jordan:** I want your job!

3 Work in pairs. Practice the conversation.

4 Complete the sentences about you. Share your information with a partner.

I sometimes _____ at lunchtime.

I usually _____ in the morning.

Today I'm _____.

**Now I can...** talk about what people do.
☐ Not at all ☐ Well ☐ Very well

**VOCABULARY: Jobs**

1. Listen and repeat.

1. a server _____
2. a flight attendant _____
3. a chef _____

4. a mechanic _____
5. a store clerk _____
6. a doctor _____

 Online Practice

2. Match each task with a job in Activity 1. Write the correct letter.

   a. repairs cars
   b. brings food
   c. helps sick people
   d. helps airplane passengers
   e. prepares food
   f. sells things

 3. Think of more jobs. What does each person do? Share your ideas with a partner.

**Pronunciation: Word stress**

1. Listen and repeat.

   | first syllable | second syllable |
   |---|---|
   | • | • |
   | <u>man</u>ager | as<u>sis</u>tant |
   | _____ | _____ |
   | _____ | _____ |
   | _____ | _____ |

2. Write each word in the correct column above.

   repair    doctor    passenger    attendant    mechanic    server

 3. Listen and repeat. Were your answers to Part 2 correct?

> > > > > **Now I can...** talk about jobs.
☐ Not at all  ☐ Well  ☐ Very well

**UNIT 8** | I love my work.   49

# GRAMMAR: Present simple and present continuous

| Present simple and present continuous | |
|---|---|
| **We use the present simple for regular activities.** | **We use the present continuous for activities happening now.** |
| She usually **prepares** vegetables in the morning. | He**'s repairing** a car right now. |
| He **works** in the evening. | They**'re watching** baseball on TV. |

Grammar Reference page 123

Online Practice

1  Look at the conversation on page 48. Circle the verbs in the present simple. Underline the verbs in the present continuous.

2  Listen to the affirmative statements. Write negative sentences with the present continuous.

   **Example:**

   You hear:   *He repairs computers.*

   You write:  *He isn't repairing computers now.*

3  Complete the sentences. Use the present simple or the present continuous form of the verb.

   1. Abby ____runs____ (run) every morning. Right now, she ____'s waiting____ (wait) for her friends at the park. She _____ (talk) to Mike.
   2. Maria _____ (drive) to the station today. She usually _____ (walk), but it _____ (rain) today.
   3. Yoshi usually _____ (take) a walk in the evening, but this evening he _____ (watch) soccer on TV.
   4. Tim and John are computer engineers. They _____ (repair) computers. They _____ (not repair) computers right now. They _____ (play) a computer game.
   5. Emma and Daniel _____ (work) in the same office. Right now, they _____ (take) a break.

4  Write sentences about yourself. What are you doing now? What do you usually do? Use both present simple and present continuous sentences. Use expressions from the box or your own ideas.

| go to English class | watch TV | see my friends | use a computer |
| go shopping | study | work | exercise |

   **Example:**

   *I am studying English now. I usually see my friends after school.*

> > > > **Now I can...** use the present simple and the present continuous.
   ☐ Not at all   ☐ Well   ☐ Very well

50

# D  READING AND WRITING

**1** Read and listen. Who are the people? Where are they from?

## On the Job

My name's Kemal, and I live in Turkey. I'm a **bus driver**. I take groups of **tourists** to visit palaces, castles, markets, and places like that. I sometimes take people from Turkey to other countries, too. I'm in Istanbul today with a group of visitors from Japan. They're visiting the
5 Topkapi Palace. Right now, I'm taking a break because they're **looking around** the palace with a tour guide. I'm drinking a coffee and reading the newspaper.
   I enjoy my job. I go to a lot of places and I meet people from other countries. I'm **away from home** a lot, so I only see my family and
10 friends on weekends.

My name's Eliana, and I'm a **real estate agent** on the coast of Brazil. I sell houses and apartments. Most of my **clients** are from big cities. They buy vacation homes here because of the beautiful beaches. Some people come here to live when they **retire**, too.
15 I'm looking at an apartment today because the owners are selling it. Right now, I'm measuring the rooms and taking some photographs.
   This is a nice place to live, and I like my job. I meet a lot of interesting people, but there are some problems, too. I often spend two or three days with **clients**. I take them to lots of houses and
20 apartments, but then they don't buy anything.

Online Practice

**2** Choose the best answer for each question.

1. Where does Kemal take the tourists?
   a. other countries
   b. famous places in Turkey
   c. only Topkapi Palace

2. When does Kemal see his family and friends?
   a. on weekdays
   b. on vacations
   c. on weekends

3. Why do most of Eliana's clients buy vacation homes on the coast of Brazil?
   a. because of the beautiful beaches
   b. because they're retired
   c. because they like big cities

4. What is the problem Eliana has with her job?
   a. She is too busy.
   b. She sees too many clients.
   c. Many clients don't buy anything.

**3** Imagine you are at work now. Write about your job (your real job or a job you want).

My name's _____, and I'm from _____. I'm a/an _____ (job). In my job, I _____. Right now, I'm _____. I like my job because _____. There are some problems, too. For example, _____.

> > > > **Now I can...** write about a job.
☐ Not at all  ☐ Well  ☐ Very well

UNIT 8 | I love my work.   51

## E  YOUR STORY: Peter decorates his home.

**1** Read and listen to the story. What is Peter working on?

**Ryan:** Hi, Peter. What's new?
**Peter:** Not much.
**Ryan:** How does Jordan's apartment look?
**Peter:** Great! In fact, I like it so much I'm doing some home decorating myself.

**Ryan:** Oh?
**Peter:** I'm putting together a new table for the living room.
**Ryan:** I'm pretty good at that kind of thing.
**Peter:** I need a little help. Could you come over this afternoon? And bring some tools?
**Ryan:** Of course.

**Ryan:** Wow! You do need some help.
**Peter:** Yeah. I'm having some trouble.

**Peter:** Could I borrow your hammer, please?
**Ryan:** Here you are.
**Peter:** Can I also have a nail?
**Ryan:** Help yourself.
**Peter:** Could you hold the nail, please?
**Ryan:** I don't think so.

**2** Listen again and answer the questions.

1. Why is Peter doing some home decorating?
2. Why does Ryan go to Peter's apartment?
3. How is Peter doing on his home decorating?
4. Why doesn't Ryan want to hold the nail for Peter?
5. What kind of friend is Ryan? Why do you think so?

**3** Use the story to complete the expressions in the box. Listen and check.

**Everyday expressions—Polite requests**

| Requests | Responses |
|---|---|
| Can/Could I _____ your hammer, please? | _____ course. |
| Can/Could you hold the nail, _____? | _____ yourself. |

**4** Work with a partner. Talk about the story. Who are you more like, Peter or Ryan? Why?

> > > > **Now I can...** make and respond to polite requests.
☐ Not at all   ☐ Well   ☐ Very well

## F REAL-WORLD LISTENING: Mountain pilots

1 Look at the picture and the map. What do you think people use helicopters for in the Alps?

2 Complete the sentences about helicopter pilots in the Alps. Use the words in the box.

| tourists | stress | Austria | food | dangerous | training |

Johannes Scheffel is a helicopter pilot in _____. He takes _____ to hotels in the mountains. He doesn't take _____. Pilots need a lot of _____ for this job. It is _____ work with a lot of _____.

3 Watch or listen and check your answers.

4 Watch or listen again. Mark the sentences true (**T**), false (**F**), or no information (**N**).

____ 1. Johannes can repair helicopters.
____ 2. His helicopter can go 150 miles per hour.
____ 3. Today, he is taking food to a home.
____ 4. Some hotels get food orders by train.
____ 5. Johannes can fly 40 trips in a day.
____ 6. Pilot training takes 1,000 days.

5 Which of these people do you think has the best job? Why? Tell a partner.
- a helicopter pilot in the Alps
- a hotel manager in the Alps
- a trip scheduler for the helicopter company

### YOUR NETWORK

**IN CLASS:** Tell a partner about your dream job. Why do you want this job?

**ONLINE:** Tell a partner about someone from your social network. What is his/her dream job? Why does he/she want this job? You can share a picture of this person.

> > > > **Now I can...** talk about a job in a different country.
☐ Not at all  ☐ Well  ☐ Very well

**UNIT 8** | I love my work.

# REVIEW Units 5-8

Circle the correct word or words to complete each sentence.

## A | Vocabulary

1. I exercise at a *gym / gas station / public restroom*.
2. I need some money. Where is a *hotel / supermarket / bank*?
3. We have a new *sofa / stove / toilet* in the kitchen.
4. My dresser is in the *kitchen / bathroom / bedroom*.
5. The ninth month is *May / September / December*.
6. March is the *first / second / third* month of the year.
7. A *salesclerk / server / mechanic* repairs cars.
8. A doctor *helps sick people / brings food / sells things*.

## B | Grammar

1. There *is / are* two gas stations in town.
2. *Is / Are* there a bank near here?
3. A: Are there any restaurants near here?
   B: Yes, *there is / there are*. There's one on 6th Street.
4. Mark *am watching / is watching / are watching* TV.
5. They *am not cleaning / isn't cleaning / aren't cleaning* the house.
6. A: Are you sleeping?
   B: *No, I'm not. / Yes, I are. / No, you aren't.*
7. Lisa *can swim / cans swim / can swims* very well.
8. You *can't play / can play not / no can play* soccer now.
9. *Cans she speak / Speak she can / Can she speak* French?
10. We *are talking / talk* on the phone every day.
11. It *is raining / rains* now. Don't go outside.
12. I *don't usually work / am not usually* working in the afternoon.

## Now I can...

talk about places around town (page 31)

talk about rooms and furniture (page 37)

talk about dates (page 43)

talk about jobs (page 49)

describe a town with *there is* and *there are* (page 32)

talk about what people are doing (page 38)

use *can/can't* for ability, possibility, and permission (page 44)

use the present simple and the present continuous (page 50)

## C | Reading

1. Read and listen to the story. What activities is Emma doing today? Check ☑ the correct picture(s).

# Home **Sweet** Home

Emma lives in Miami. She has a small apartment. It only has one bedroom, and the kitchen is very small, but it has a great view. It's on the 40th floor, so she can see the ocean and
5 most of downtown from her window. There's a park right across the street. It is her first apartment. She doesn't have much furniture—only a bed, a table, two chairs, and a sofa. Emma is from Costa Rica, and her family still lives there.
10 She can't visit them often, but she talks to them every week.

Emma really likes Miami. She works at a restaurant in South Beach. South Beach has many good restaurants. It's a fun place to work.
15 Emma is a server, but she wants to be a chef. She is saving money to take cooking classes.

Today is Monday, so Emma isn't working. It's her day off. On Mondays, she usually goes to the gym for a yoga class. Then she sometimes
20 goes shopping or cleans her apartment. Today she isn't doing yoga. It's her birthday, so she's having lunch with her friends.

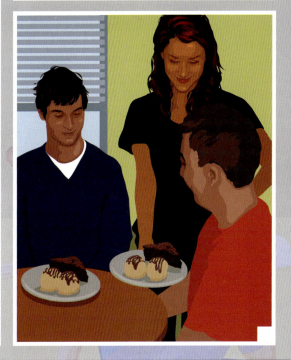

2. Read the story again. Answer the questions about Emma.
   1. What are the good things about Emma's apartment?
   2. What are the bad things about Emma's apartment?
   3. Where does Emma's family live?
   4. Why does Emma want to take cooking classes?
   5. What does Emma usually do on Mondays?
   6. What is Emma doing today? Why?

# GET CONNECTED

## Getting started online

**Looking** When people join an online group or social network, everything is new. First, they open an account and create a profile. A profile is the first way people introduce themselves. Then, they look around. In some social networks, such as Facebook, people find "friends." In other social networks, such as Twitter or Tumblr, users "follow" other users.

**Learning** Soon users understand how the group works. They learn the "rules." They learn about the other people in the network, and those people share information. People can share information in many ways. For example, users can learn about a topic through a blog or an online video.

**Participating** Most users look and learn about a network before they participate. When you participate, you "talk" to other people in the network. You can post answers to other people's comments. You can give your opinion. You can post links to videos, music, or online articles.

### GET Started

**1** **Look** at the Web page for Cozy Cup cafe on page 57. What information do you see? Mark the following statements true (**T**) or false (**F**).

_____ 1. Only the cafe gives information on this page.

_____ 2. Customers can write on this page.

_____ 3. You can read the menu here.

_____ 4. The same person makes all the comments.

_____ 5. You can find out about events at Cozy Cup here.

**2** Read the posts and comments. What are three things you **learned** about Cozy Cup?

1. _____
2. _____
3. _____

## Cozy Cup  Restaurant, New York City

**About:** Cozy Cup is a cafe in New York. It is a family business. Ryan and Cindy Gaskell are the owners.

### Posts

 **Cozy Cup** Good morning, pastry fans! Today we have blueberry and apple muffins. 10% discount with a large coffee. 1 hour ago
Share    👍 12 people like this

**Lucy Patterson** I love the blueberry muffins! See you soon.
40 minutes ago

**Hassan Nasr** My favorite muffin is apple. 37 minutes ago

**John Williams** Those muffins are terrible, and the coffee is like dirty water. Try another restaurant. 15 minutes ago

 **Cozy Cup** Enjoy delicious food and drinks and listen to a local poet read. This afternoon at 4:00, Max Brodsky reads from his new book, *A Summer Sky*. Wednesday, 10:00 a.m.
Share    👍 8 people like this

**Jackie Okimba** His reading is wonderful! I also write poetry. I can come and read sometime. I think everyone will like it.
Wednesday, 10:30 a.m.

### Photo Album

### You and Cozy Cup

**Michael Coffey** What a great cafe! A friendly place and amazing food!

**Lisa T. Smith** All I can say is I love the coffee. Yum.

Write a comment    **Post**

---

**3** **Look** at the Web page again for possible "rules" for this page. Check ✓ the ones you see.

☐ 1. The cafe can post announcements about the menu.
☐ 2. The cafe can post announcements about events.
☐ 3. People can post opinions about the menu.
☐ 4. The cafe answers people's comments.
☐ 5. People can post pictures of the cafe.

**4** Work with a partner. Which comments on the page do you think are good? Which do you think are not good? Why? What do you learn about the people from their comments?

**5** Try **participating**. Write a comment for one of Cozy Cup's announcements.

Leave a comment
Name:
Comment:

**> > > > Now I can...** post comments online.
☐ Not at all   ☐ Well   ☐ Very well

**Take it online**
Visit the website of a neighborhood restaurant.

57

# UNIT 9 Where were you?

**YOUR NETWORK**

Go to Network Online Practice to record your voice in the conversations on pages 58 and 62.

Go to Network Online Practice to watch video about Memphis, an American city.

Network! Go online to find someone from another country who went to a fun place last weekend. Share on page 63.

Online Practice

### A CONVERSATION: We were at a play.

1. Look at the picture. Where are they?

2. Read and listen.
   CD 2-23

   **Sarah:** Were you at home last night?
   **Ryan:** No, we weren't. We were at a play.
   **Sarah:** Was it good?
   **Ryan:** No, it wasn't. It was boring, and the actors weren't very good.
   **Sarah:** What play? Was it *The Orange Tree*?
   **Ryan:** Yes.
   **Sarah:** Oh, no. We have tickets for that play tonight!

3. Work in pairs. Practice the conversation.

4. Stand and ask your classmates about where they were last night.

   Were you at home last night?
   Yes, I was./No, I wasn't.

   What did you do?
   Where were you?

**Now I can...** say where I was last night.
☐ Not at all  ☐ Well  ☐ Very well

## B VOCABULARY: Events and places to go

1  Listen and repeat.

1. the theater / a play

2. a concert

3. the movies

4. a shopping mall

5. a soccer game

6. an art gallery

Online Practice

2  What other events and places do you know? Share your ideas with a partner.

_____  _____

_____  _____

> **Language note** *to/at*
> They're going **to** a concert.
> They're **at** a concert.

3  Listen. Where are the people?

**Example:**

*They're at a soccer game.*

4  Talk to a partner. How often do you go to the places and events in Activity 1? Use *often*, *sometimes*, or *never* in your answer.

> How often do you go to the theater?

> I sometimes go to the theater.

## > > > > > Now I can... talk about places to go.
☐ Not at all  ☐ Well  ☐ Very well

UNIT 9 | Where were you?

## GRAMMAR: Past simple—*to be*

Grammar Reference page 124

| Affirmative statements | Negative statements | Questions and short answers | |
|---|---|---|---|
| I **was** at the movies last night. She **was** at home yesterday. | We **weren't** at a soccer game on Saturday. They **weren't** away last week. | **Were** you away last week? | Yes, we **were**. No, we **weren't**. |
| | | **Was** the play good? | Yes, it **was**. No, it **wasn't**. |

Online Practice

1. Look at the conversation on page 58. Circle *was*, *wasn't*, *were*, and *weren't*.

2. Complete the conversation with *was*, *wasn't*, *were*, or *weren't*.

**Jen:** ___Were___(1) you away last week?

**Owen:** Yes, I was. I _____(2) on vacation with some friends.

**Jen:** Oh, _____(3) you at the beach?

**Owen:** No, we _____(4). We _____(5) in Dubai. It _____(6) great.

**Jen:** Yes, I _____(7) there last year. How was the vacation?

**Owen:** Well, the hotel _____(8) excellent.

**Jen:** _____(9) the weather good?

**Owen:** Yes, it _____(10). Was it OK here?

**Jen:** No, it _____(11). It _____(12) cold and rainy. I _____(13) at home all weekend.

Burj Al Arab

3. Practice the conversation in Activity 2 with a partner.

### Pronunciation: *was/wasn't; were/weren't*

**1 Listen and repeat.**

1. We were away last week.   We weren't away last week.
2. The food was very good.   The food wasn't very good.
3. I was here yesterday.   I wasn't here yesterday.
4. They were at the mall.   They weren't at the mall.

**2 Listen. Which sentence in each pair above do you hear? Circle it.**

>>>> **Now I can...** say where people were.
☐ Not at all   ☐ Well   ☐ Very well

# D  READING AND SPEAKING

1 Read and listen. Did Teresa and Marco like their hotels?

CD 2-29

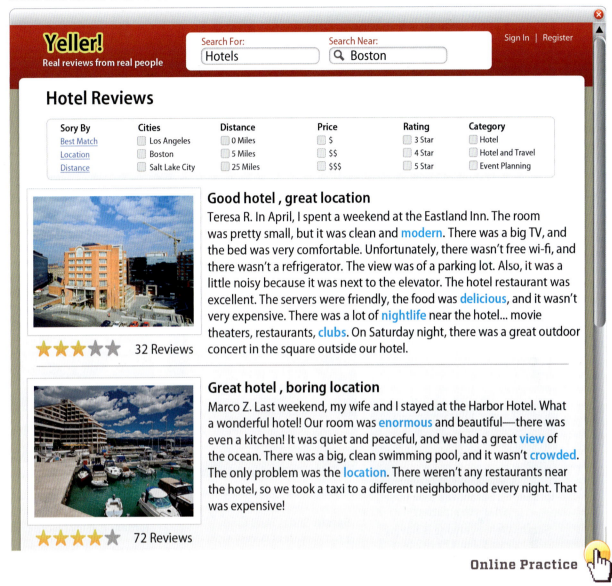

**Yeller!** Real reviews from real people

Search For: Hotels
Search Near: Boston
Sign In | Register

## Hotel Reviews

| Sort By | Cities | Distance | Price | Rating | Category |
|---|---|---|---|---|---|
| Best Match | ☐ Los Angeles | ☐ 0 Miles | ☐ $ | ☐ 3 Star | ☐ Hotel |
| Location | ☐ Boston | ☐ 5 Miles | ☐ $$ | ☐ 4 Star | ☐ Hotel and Travel |
| Distance | ☐ Salt Lake City | ☐ 25 Miles | ☐ $$$ | ☐ 5 Star | ☐ Event Planning |

### Good hotel , great location
Teresa R. In April, I spent a weekend at the Eastland Inn. The room was pretty small, but it was clean and *modern*. There was a big TV, and the bed was very comfortable. Unfortunately, there wasn't free wi-fi, and there wasn't a refrigerator. The view was of a parking lot. Also, it was a little noisy because it was next to the elevator. The hotel restaurant was excellent. The servers were friendly, the food was *delicious*, and it wasn't very expensive. There was a lot of *nightlife* near the hotel... movie theaters, restaurants, *clubs*. On Saturday night, there was a great outdoor concert in the square outside our hotel.

★★★☆☆ 32 Reviews

### Great hotel , boring location
Marco Z. Last weekend, my wife and I stayed at the Harbor Hotel. What a wonderful hotel! Our room was *enormous* and beautiful—there was even a kitchen! It was quiet and peaceful, and we had a great *view* of the ocean. There was a big, clean swimming pool, and it wasn't *crowded*. The only problem was the *location*. There weren't any restaurants near the hotel, so we took a taxi to a different neighborhood every night. That was expensive!

★★★★☆ 72 Reviews

Online Practice

2 What were the good things and bad things about each hotel? Complete the chart.

|  | Good | Bad |
|---|---|---|
| Eastland Inn |  |  |
| Harbor Hotel |  |  |

  3 Talk with a partner. Which hotel do you like better? Why?

  4 Work with a partner. Write a conversation between Teresa, Marco, and their friends. Then practice the conversation.

Friend: *How was the Eastland Inn?*

Teresa: *It was pretty good. The room was…*

Friend: *Were the restaurants…?*

> > > > > **Now I can...** say how good or bad something was.
☐ Not at all ☐ Well ☐ Very well

UNIT 9 | Where were you?

## E  YOUR STORY: A bad news day

1  Read and listen. Who has bad news?

**1**

**Sarah:** Hi, Lucy. How was the interview?
**Lucy:** Not so good. There wasn't a position available after all.
**Sarah:** Why not?
**Lucy:** The personal assistant was on sick leave, but now she's back.
**Sarah:** Oh, I'm sorry to hear that.

**2**

**Sarah:** Hey, do you know about the Jobs Group?
**Lucy:** No, I don't. What is it?
**Sarah:** I can send you the website. The group helps people use networks to learn about jobs.
**Lucy:** Great. Thanks. See you later.

**3**

**Jordan:** Hi, Sarah. You look upset. What's the matter?
**Sarah:** Well, everyone is having a bad day. I have a test tomorrow, and Lucy didn't get the job.

**4**

**Jordan:** Oh, no. She must be so upset.
**Sarah:** She is. And Cindy's mother is very sick.
**Jordan:** That's terrible.
**Sarah:** Yes, she's very worried.

**5**

**Jordan:** Lucy! Are you OK?
**Lucy:** Now I am. I went to get some ice cream.
**Jordan:** Is it my favorite, double-chocolate fudge?
**Lucy:** Yes, it is. Or actually, it was.

Online Practice

2  Listen again. Write true (**T**), false (**F**), or no information (**N**).

_____ 1. Lucy didn't get the job because she had a bad interview.
_____ 2. Sarah tells Lucy about a group for finding jobs.
_____ 3. Cindy's mother is in the hospital.
_____ 4. Jordan is having a bad day.
_____ 5. Jordan is looking for Lucy.
_____ 6. Jordan wants to eat ice cream.

3  Use the story to complete the expressions in the box. Listen and check.

4  Work in a group. Talk about a time you had bad news.

**Everyday expressions—Talking about bad news**

I'm sorry to _____ that.
What's the _____?
That's _____.

> > > **Now I can...** talk about bad news.   > > >
☐ Not at all   ☐ Well   ☐ Very well

## F REAL-WORLD LISTENING: An American city

1. Look at the photo and map. What do you know about Memphis? What do you think it is famous for?

2. Watch or listen to the report about Memphis. What kinds of music does the narrator mention? Where can you hear this music?

Online Practice

3. Watch or listen again. Match the place with the correct description.

____ 1. Sun Studio          a. There are over 25 clubs here.
____ 2. Beale Street         b. This was Elvis Presley's house.
____ 3. Peabody Hotel        c. Rock and roll was born here.
____ 4. Graceland            d. This place is now a museum.
____ 5. Lorraine Motel       e. This place is in an ideal location.

4. Imagine your partner is visiting your hometown. What places are good to visit? What can you do and see there? Tell a partner.

| Places | Things to do and see |
|--------|----------------------|
|        |                      |
|        |                      |
|        |                      |

### YOUR NETWORK

**IN CLASS:** Interview three classmates. Where was each person last weekend? What did he/she do there?

**ONLINE:** Tell a partner about someone from your social network. What country is the person from? What fun place did he/she go last weekend? What did he/she do there? You can share a picture of this person.

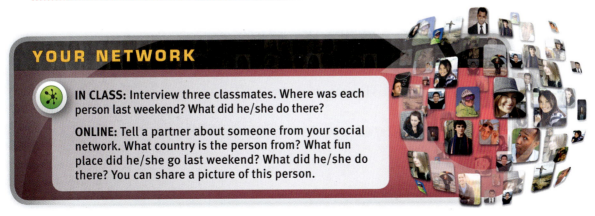

> > > > > **Now I can...** talk about things to do and see in a city.
☐ Not at all  ☐ Well  ☐ Very well

# UNIT 10 What did you study?

## YOUR NETWORK

Go to Network Online Practice to record your voice in the conversations on pages 64 and 68.

Go to Network Online Practice to watch video about a high school in the United States.

Network! Go online to learn about someone's favorite school subjects. Share on page 69.

Online Practice

### A CONVERSATION: What was your favorite subject?

1 Look at the picture. What do you think Peter is doing?

2 Read and listen.
CD 2-33

**Cindy:** Did you go to college, Peter?
**Peter:** Yes, I did. I went to New York University.
**Cindy:** What was your favorite subject?
**Peter:** I liked business. It was my favorite.
**Cindy:** Really? What did you like about it?
**Peter:** I was good at it. And I like math.
**Cindy:** I liked English, but I didn't finish college.
**Peter:** Why not?
**Cindy:** I needed a job.

3 Work in pairs. Practice the conversation.

4 Stand and ask your classmates about their favorite subjects.

What is/was your favorite subject?   What do/did you like about it? Why?

I like/liked...

**Now I can...** talk about my favorite school subjects.
☐ Not at all   ☐ Well   ☐ Very well

## B VOCABULARY: School subjects

**1** Listen and repeat.

1. foreign languages

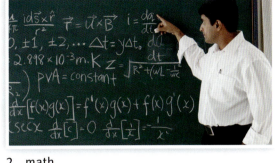
2. math

3. physical education (P.E.)

4. art

5. science

6. history

**Online Practice**

**2** Write more school subjects.

1. _____  3. _____
2. _____  4. _____

**3** Listen. Linda and her grandchildren are talking about school. What subjects do they like? What subjects don't they like? Complete the chart.

|  | Adam | Holly | Joe |
|---|---|---|---|
| Likes |  |  |  |
| Doesn't like |  |  |  |

**4** Listen again. Check ✓ the languages Linda talks about.

☐ English  ☐ French  ☐ Spanish
☐ Portuguese  ☐ Chinese  ☐ Japanese

> > > > > **Now I can...** talk about school subjects.
☐ Not at all  ☐ Well  ☐ Very well

UNIT 10 | What did you study? 65

## C GRAMMAR: Past simple—statements

### Grammar Reference page 124

| Regular verbs | |
|---|---|
| finish | I **finished** college. |
| need | She **needed** a job. |
| like | They **liked** art. |

| Negative statements | |
|---|---|
| He **didn't like** it. | NOT He didn't liked it. |
| I **didn't have** a job. | NOT I didn't had a job. |

| Irregular verbs | |
|---|---|
| get up | I **got up** at 7:00 today. |
| go | He **went** to school yesterday. |
| have | We **had** a party last night. |
| take | They **took** a test this morning. |
| teach | She **taught** math for three years. |

Online Practice

1. Look at the conversation on page 64. Underline the regular past tense verbs. Circle the irregular past tense verbs.

2. Complete the sentences with the past tense form of a verb from the box.

> get up    go    eat    take    like    study

   1. They _____ to the beach last weekend.
   2. We _____ dinner at a restaurant last night.
   3. I _____ a Spanish class last year.
   4. Nat _____ at 9:00 yesterday.
   5. I _____ math for eight years.
   6. Carmen _____ her history teacher.

3. Did you do these things yesterday? Write sentences.
   **Example:**
   *I went to work.* OR *I didn't go to work.*
   1. go to work _____
   2. get up before 7:00 a.m. _____
   3. play a computer game _____
   4. eat at a restaurant _____
   5. have a party _____
   6. watch TV _____

 4. Work with a partner. Talk about what you did last summer.
   **Example:**
   *Last summer, I visited my grandparents in Mexico. I also went to the beach with my friends.*

### Pronunciation: Past tense /ɪd/ endings

1. Which verbs have an extra syllable /ɪd/ in the past tense? Check ✓ them.
   - ☐ 1. like    liked
   - ☐ 2. want    wanted
   - ☐ 3. start    started
   - ☐ 4. work    worked
   - ☐ 5. study    studied
   - ☐ 6. need    needed

2. Listen and repeat. Were your answers to Part 1 correct?

>>>>> **Now I can...** talk about the past.
☐ Not at all   ☐ Well   ☐ Very well

# D. READING AND WRITING

**1** Read and listen. What was Eric's problem?

## The New Eric

Eric Johnson was a happy child. He lived in a small town with his parents, his brother, and his sister. He was a **straight-A** student, and his favorite subjects were math and science. In his free time, he loved to play video games and play the piano.

After college, Eric got a job as a **video game designer**. Soon, he got married and had three children. When he was 32, he had a terrible **accident** and **injured** his head. He didn't **remember** anything about his life before the accident. He didn't know his friends or family—not even his wife. He couldn't read or do math. Fortunately, he could still do one thing—play the piano!

Eric got a **tutor**. His tutor taught him math, science, reading, and history—the same subjects he learned as a child. It wasn't easy, but Eric worked hard. His children helped him with his homework. Again, he was an excellent student. And again, his favorite subjects were math and science. Later, he took high school classes online. After he finished high school, he went to college for the second time! On the first day, he was very **nervous**. But the other students were friendly, and he soon felt comfortable. Like "old Eric," "new Eric" decided to design video games. Now he has a good job and is very happy again!

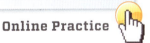

Online Practice

**2** Number the events in the correct order (1 to 5).
_____ a. Eric learned math and science from his tutor.
_____ b. Eric had a terrible accident and injured his head.
_____ c. Eric was very nervous on the first day of college.
_____ d. Eric got married and had three children.
_____ e. Eric took high school classes online.

> **Language note:** *could*
> *Could* is the past tense of *can*.
> *Couldn't* is the past tense of *can't*.

**3** Complete these sentences about your education.
1. I started school in _____.
2. The name of the school was _____.
3. On my first day, I _____.
4. I (liked/didn't like) the school because _____.
5. My favorite subject was _____.
6. I didn't like _____.

> > > > > **Now I can...** write about my education.
☐ Not at all  ☐ Well  ☐ Very well

## E YOUR STORY: Lucy meets Jordan's brother.

1 Read and listen. Who is with Jordan?

**Jordan:** Hi, Lucy. This is my brother Matt.
**Lucy:** How wonderful! I didn't know your brother was visiting.
**Jordan:** It was a surprise to me, too.
**Lucy:** Well, Matt, it's very nice to meet you.
**Matt:** Same here. Jordan talks about you all the time.
**Lucy:** Oh, he does? I hope it's all good.
**Matt:** It is.

**Lucy:** It's so nice you can be closer to Jordan.
**Jordan:** Matt just graduated from college in Australia.
**Lucy:** Congratulations! That's terrific! What did you study?
**Matt:** Physical education. I like sports a lot.
**Jordan:** And he played a lot of computer games.

**Lucy:** Do you know much about computers?
**Matt:** Yes, I do. Do you need help with something?
**Lucy:** Yes. I'm trying to find a new job. Do you think social networks can help?
**Matt:** Maybe. Let me show you a few networking sites.

**Lucy:** Jordan, your brother is so nice. He's helping me a lot.
**Jordan:** That's good.
**Lucy:** And he's so friendly and fun.
**Jordan:** That's what all the girls say.

Online Practice

2 Listen again. Answer the questions.

1. Did Matt tell Jordan about his visit?
2. What is Matt's good news?
3. What did Matt study in college?
4. What is Lucy doing on her computer?
5. How does Matt help Lucy?
6. What does Lucy think about Matt?

3 Use the story to complete the expressions in the box. Listen and check.

4 Work with a partner. Share some good news you had in the past year.

**Everyday expressions—Talking about good news**

How _____!
Congratulations!
That's _____!
That's _____.

> > > > > **Now I can...** talk about good news.
☐ Not at all  ☐ Well  ☐ Very well

68

##  REAL-WORLD LISTENING: Hawthorne High

1. Look at the picture. Does this look like your school or any schools in your city?

2. Watch or listen to the report about Hawthorne High. What is Alison's favorite subject? What are two subjects she likes and two she doesn't like?

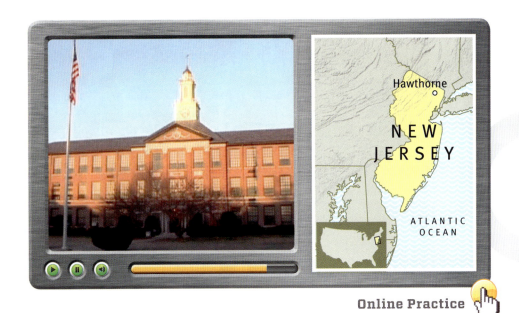

Online Practice

3. Watch or listen again. Match these words from the video with the definitions.

____ 1. senior          a. part of a class schedule
____ 2. hang out        b. physical education
____ 3. chat            c. student in the last year of high school or college
____ 4. period          d. have a friendly conversation
____ 5. awesome         e. not delicious
____ 6. gym             f. spend time in a place or with someone
____ 7. be into         g. cool; a word for something you really like
____ 8. gross           h. enjoy or be interested in something

4. Tell a partner about your high school. What classes did you take?
What time did school start and finish?
What did you like and not like?

### YOUR NETWORK

**IN CLASS:** Interview three classmates. What are/were their favorite school subjects? Why do/did they like them?

**ONLINE:** Tell a partner about someone from your social network. What are/were his or her favorite school subjects? Are they the same as yours? You can share a picture of this person.

> > > > **Now I can...** talk about my school.
☐ Not at all  ☐ Well  ☐ Very well

# UNIT 11 What happened to you?

## YOUR NETWORK

Go to Network Online Practice to record your voice in the conversations on pages 70 and 74.

Go to Network Online Practice to watch video about about how laughter is the best medicine.

Network! Go online to find someone who had an accident. Share on page 75.

Online Practice

### A CONVERSATION: I hurt my knee.

1. Look at the picture. What do you think happened?

2. Read and listen.
CD 2-43

**Peter:** Hi, Henry. What happened to you?
**Henry:** Well, I went skiing last weekend…
**Peter:** Oh, did you break your leg?
**Henry:** No, but I hurt my knee.
**Peter:** How did you do that? Did you have a bad skiing accident?
**Henry:** No, I didn't. I fell out of bed.

3. Work in pairs. Practice the conversation.

4. Stand and ask your classmates about what they did last week. Try to ask more questions.

> What did you do last weekend?
>
> Well, I went _____ last weekend.
>
> Really? Was it fun?
>
> Yes, it was./No, it wasn't.

**Now I can...** talk about past events.
☐ Not at all  ☐ Well  ☐ Very well

## B VOCABULARY: Parts of the body

1  Write the missing parts of the body. Use the words in the box.

| hand | arm | eye | leg | head | nose |

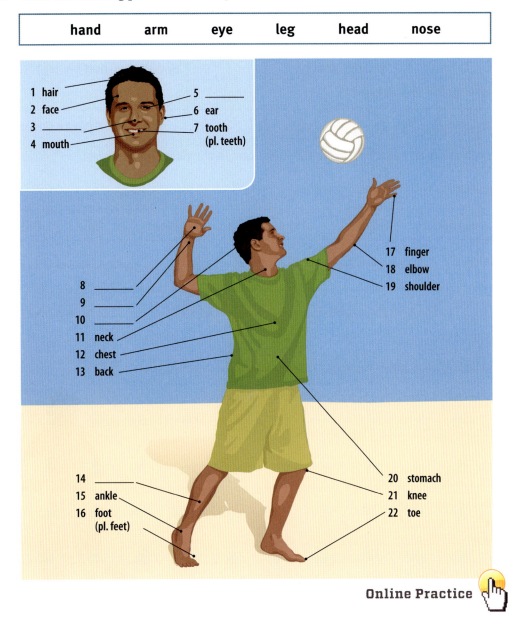

 2  Listen and repeat.

CD 2-44

3  Read the sentence. Write the body part.

1. You speak with this.   *It's your mouth.*
2. These are at the end of your feet.
3. This is between your eyes and your mouth.
4. This is between your leg and your foot.
5. You see with this part of your body.
6. This is between your head and your chest.

### Pronunciation: Silent letters

1. Circle the silent letters (the letters you don't hear).

   (k)nee     answer     stomach     who
   walk       write      know        when

 2. Listen and repeat. Were your answers to Part 1 correct?

CD 2-45

> > > > > **Now I can...** name parts of the body.
☐ Not at all  ☐ Well  ☐ Very well

## GRAMMAR: Past simple—questions

| yes/no questions | | |
|---|---|---|
| Did you go skiing? | Yes, I did. | |
| Did they enjoy it? | No, they didn't. | |
| NOT Did you went skiing? Did they enjoyed it? | | |

| Grammar Reference page 125 |
|---|
| Wh- questions |
| Where did you stay? |
| How did she hurt her leg? |

Online Practice

1  Look at the conversation on page 70. Circle the questions in the past simple tense.

2  Complete the sentences with the correct verbs.

| burn | burned | break | broke | go | went | fall | fell |

1. A: How did you ___burn___ your hand?
   B: I _____ it on the stove.
2. A: I _____ to the hospital yesterday.
   B: Oh, why did you _____ there?
3. A: I _____ my leg when I was a child.
   B: Oh, how did you _____ it?
4. A: Did Neil _____ off his bike?
   B: No, he didn't. He _____ off a chair.

3  Complete the questions and short answers. Use the words in parentheses.

**Hannah:** ___Did you have___ (you / have) a good weekend?

**Carlos:** ___Yes, I did___ (Yes), thanks.

**Hannah:** What _____ (you / do)?

**Carlos:** I went mountain biking.

**Hannah:** Where _____ (you / go)?

**Carlos:** The White Mountains in New Hampshire.

**Hannah:** _____ (you / go) with Lena?

**Carlos:** _____ (No). I went with some guys from work.

**Hannah:** _____ (you / enjoy) it?

**Carlos:** _____ (Yes), thanks.

4  Listen. Were your answers to Activity 3 correct? Practice the conversation with your partner.

5  Talk with a partner about your weekend. Make questions with the words below.
- What / do?
- Who / go with?
- Where / go?
- Have / a good time?

> > > > **Now I can...** ask questions about past events.
☐ Not at all  ☐ Well  ☐ Very well

#  READING AND SPEAKING

1 Read and listen. What happened to Nina and John?

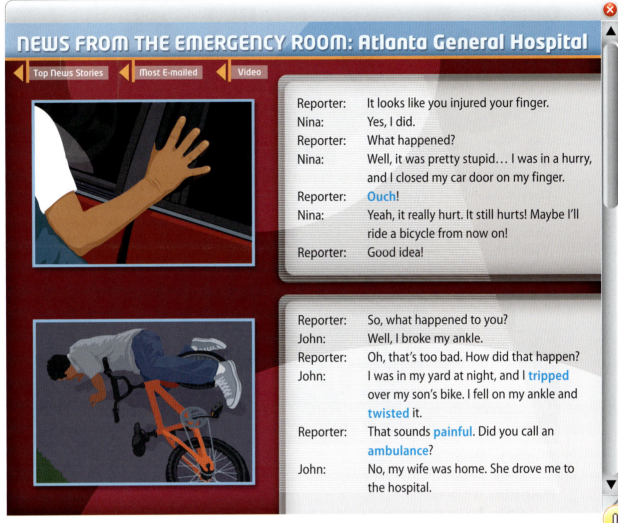

2 Answer the questions.
1. How did each person get hurt?
2. What body part did each person injure?
3. Why did each person talk about a bicycle?
4. Which injury do you think was the most painful?

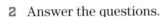 3 Interview a partner about an accident that he/she had. Use the questions in the box.

| What happened? | How much did it hurt? | Who helped you? |
| --- | --- | --- |
| Where did it happen? | When did it happen? | Did you go to the hospital? |

**Example:**
A: *What happened?*
B: *I had an accident (last year)…*

> > > > > **Now I can...** talk about an accident.

☐ Not at all ☐ Well ☐ Very well

UNIT 11 | What happened to you? 73

**YOUR STORY:** Ryan has a cold.

1. Read and listen. What is the problem?

**Cindy:** You don't look so good, Ryan.
**Ryan:** I feel terrible.
**Cindy:** What's the matter?
**Ryan:** I have a cold, my nose is stuffy, and I have a headache. Oh, and I have a cough, too.

**Cindy:** Poor baby. Do you want some hot tea?
**Ryan:** I think I need some real medicine.
**Cindy:** I guess I can go to the drugstore.
**Ryan:** Really? That's great. Oh! Could you buy some honey and lemon at the supermarket, too? My throat is a little sore. And maybe some cookies?

**Pharmacist:** Hello. Can I help you?
**Cindy:** Yes. Do you have anything for a cold?
**Pharmacist:** Is it for you?
**Cindy:** No, I'm the healthy one. It's for my husband.
**Pharmacist:** This medicine is very good. He can take two capsules every 12 hours.

**Cindy:** Here you go, Ryan. The pharmacist says to take two now, and then two in 12 hours. And drink some orange juice.
**Ryan:** Oh, thanks. Can you get me some orange juice?
**Cindy:** Sure.
**Ryan:** And how about some soup?

Online Practice

2. Listen again. Circle the answer.

1. Ryan has *a cold / the flu*.
2. He also has a *stomachache / sore throat*.
3. Cindy is going to the *drugstore / supermarket* to buy medicine.
4. Ryan asks Cindy to buy some *cookies / soup*.
5. The pharmacist says to take the medicine every *two / twelve* hours.
6. Ryan needs to drink some *orange juice / tea*.

3. Use the story to complete the expressions in the box. Listen and check.

4. Work in a group. Tell the group about medicine you take when you are sick.

**Everyday expressions—Buying medicine**

Do you have _____ for...?
Is it for _____?
Take two _____ every _____ hours.

> > > > > **Now I can...** talk about buying medicine.
☐ Not at all  ☐ Well  ☐ Very well

# REAL-WORLD LISTENING: Laughter is the best medicine.

1. Look at the photos. How often do you smile or laugh? When can laughter help people?

2. What are the benefits of laughter? Check ✓ your ideas.
   - ☐ It feels good.
   - ☐ It helps babies grow.
   - ☐ It is an international language.
   - ☐ It is good for stress.
   - ☐ It helps with pain.
   - ☐ It makes you smarter.

3. Watch or listen and check your answers.

4. Watch or listen again. Match the people with the correct description.
   ____ 1. Some doctors…      a. laugh up to 400 times a day.
   ____ 2. Scientists…         b. get better faster with laughter.
   ____ 3. Children…           c. say to laugh 30 minutes a day.
   ____ 4. Many adults…        d. help people in hospitals laugh.
   ____ 5. Patients…           e. say laughter is good for colds and the flu.
   ____ 6. Clown doctors…      f. laugh only a few times a day.

5. Imagine your partner wants to laugh more. What can he or she do? Recommend places to visit, things to do, and other ideas.

## YOUR NETWORK

**IN CLASS:** Tell a partner about a time you had an accident. When did it happen? What part of your body did you hurt?

**ONLINE:** Tell a partner about someone from your social network. What kind of accident did that person have? What part of the body did he/she hurt?

> > > > > **Now I can...** recommend ways to laugh.
☐ Not at all  ☐ Well  ☐ Very well

# UNIT 12 I'm going on a cruise.

## YOUR NETWORK

Go to Network Online Practice to record your voice in the conversations on pages 76 and 80.

Go to Network Online Practice to watch video about Honolulu, Hawaii.

Network! Go online to get some travel ideas for your next vacation. Share on page 81.

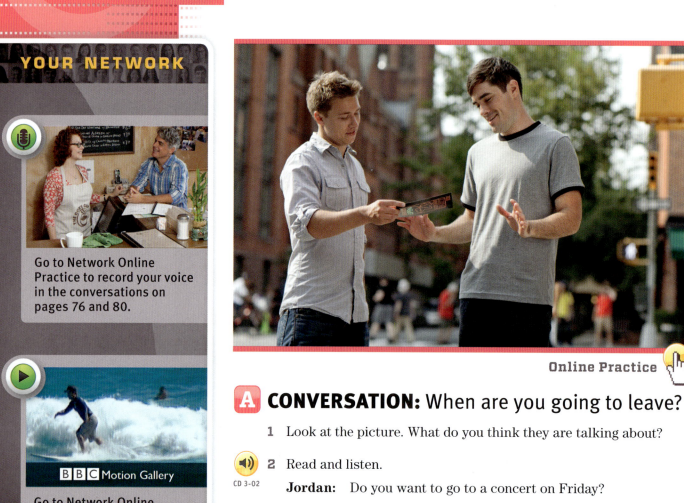

Online Practice

### A  CONVERSATION: When are you going to leave?

1  Look at the picture. What do you think they are talking about?

2  Read and listen.  
CD 3-02

**Jordan:** Do you want to go to a concert on Friday?

**Peter:** Sorry, Jordan, I can't. I'm going to go on a business trip at the end of the week.

**Jordan:** Oh. When are you going to leave?

**Peter:** On Friday morning.

**Jordan:** Wait—I think the band is going to play on Thursday, too. I'll check their website.

**Peter:** Oh, did I say Friday? I meant Thursday. Maybe next time.

3  Work in pairs. Practice the conversation.

4  Work in a small group. Invite someone in your group to an event. Try to get them to come.

Do you want to go to _____ on Friday?

I can't. I'm going to _____.

Wait—I think _____.

**Now I can...** invite someone to an event.
☐ Not at all   ☐ Well   ☐ Very well

## B VOCABULARY: Travel

1 Listen and repeat.

1. go on a business trip

2. go camping

3. go on a cruise

4. go backpacking

5. luggage

6. passport

Online Practice

2 What other travel words or phrases do you know? Share your ideas with a partner.

3 Listen to the conversation about Marco and Isabel's trip. Are the statements true (**T**) or false (**F**)?

_____ 1. They went camping.
_____ 2. They went to the Caribbean.
_____ 3. They lost their passports.
_____ 4. He left his passport at home.
_____ 5. They called their neighbor.
_____ 6. They missed the boat.

4 Work with a partner. Ask and answer the questions.

- When do you go on vacation?
- Where do you go?
- Who do you go with?
- How do you usually travel?
- Where do you stay?
- What things do you do?

**Language note:** *go* + preposition

go **on** vacation, go **on** a cruise
go **by** train, go **by** car
go **to** Brazil, go **to** the airport
BUT go **camping**, go **skiing**

### Pronunciation: Vowel sounds

1. Check ✓ the pairs with the same vowel sound.

☐ 1. trip       miss
☐ 2. lose       cruise
☐ 3. travel     wallet
☐ 4. train      plane
☐ 5. passport   camping
☐ 6. suitcase   business

2. Listen and repeat. Were your answers to Part 1 correct?

> > > > > **Now I can...** talk about travel.
☐ Not at all   ☐ Well   ☐ Very well

UNIT 12 | I'm going on a cruise.    77

## C GRAMMAR: *be + going to*

CD 3-06

**Grammar Reference** page 125

| Statements | |
|---|---|
| I'm **going to** take a vacation. | She **isn't going to** study. |
| I'm **not going to** go camping. | We're **going to** visit a museum. |
| He's **going to** relax. | They **aren't going to** have a party. |

| yes/no questions and short answers | |
|---|---|
| **Are** you **going to** study? | Yes, I **am**. |
| | No, I'm **not**. |
| **Is** she **going to** study? | Yes, she **is**. |
| | No, she **isn't**. |

Online Practice

1. Look at the conversation on page 76. Underline the statements and questions with *be + going to*.

2. What are people going to do on Saturday? Make sentences with *be + going to*. First make a negative sentence, then make an affirmative sentence. Use the words below.

   1. We / study / go shopping  *We aren't going to study. We're going to go shopping.*
   2. He / get up early / stay in bed _____
   3. They / play soccer / play tennis _____
   4. I / work in the office / relax on a beach _____

3. Check ✓ the sentences if they are true for you. If not, rewrite them so they are.
   - ☐ 1. I'm going to study after class. _____
   - ☐ 2. I'm going to go to a restaurant today. _____
   - ☐ 3. I'm going to watch TV tonight. _____
   - ☐ 4. I'm going to visit my friends this weekend. _____
   - ☐ 5. I'm going to have a party this weekend. _____

4. Ask and answer questions about the people in Activity 2.

   **Example:**

   A: *Are they going to study?*

   B: *No, they aren't. They're going to …*

5. Are you going to do these things next weekend? Ask and answer questions with a partner.

   **Example:**

   A: *Are you going to go shopping this weekend?*

   B: *Yes, I am.*

   A: *What are you going to buy?*
   *Where are you going to go?*
   *Who are you going to go with?*

   | go shopping | study | go to the movies |
   | relax | watch sports on TV | get up late |

>>>> **Now I can...** talk about my plans for the weekend.
☐ Not at all  ☐ Well  ☐ Very well

## D  READING AND WRITING

1  Read and listen. What do these people want to do after they graduate?

### Ross College Students Prepare for Future!

# Big Plans!

These people are going to graduate from college soon. They all have exciting travel plans!

**Sandy Yates (U.S.)**

What am I going to do when I graduate from college? Well, first I'm going to take it easy for a while. I'm going to go camping with my family. Then, in August, I'm going to travel to Japan. I'm going to teach English there for two years. It's going to be very different from life in the U.S.

**Hyun Soo Park (Korea)**

After college, I'm going to travel around the world with two friends. First, I'm going to work in a restaurant near my home to earn some money. Then, in October, we're going to fly to Bangkok. We're going to go backpacking in Thailand, Indonesia, and some other countries. After that, we're going to fly to Australia and look for jobs there. It's going to be great.

**Antonio Velez (Peru)**

When I finish college, I'm going to go to Canada for a year. I want to improve my English. I'm going to work in a hotel in a ski resort. That's going to be great because I love skiing. I'm not going to leave Peru until September, so I'm going to work in my parents' bookstore first.

Online Practice

2  All three people are going to travel to a different country and do something there. Write what they are going to do and in which country.

|  | Activity | Country |
| --- | --- | --- |
| Sandy |  |  |
| Hyun Soo |  |  |
| Antonio |  |  |

**Language note:** Sequencers

**First**, I'm going to work…
**Then**, we're going to…
**After that**, we're going to…

3  Imagine you're going to take a year off from your job or school. Think about your plans.

1. What are you going to do before you leave?
2. Where are you going to go?
3. What are you going to do there?
4. Who are you going to go with?
5. When are you going to leave?

4  Write a paragraph about your year off.

For my year off, I'm going to _____. First, I'm going to _____. Then, I'm going to travel to _____. I'm going to _____ there. After that, I'm going to _____. It's going to be _____.

> > > > > **Now I can...** describe plans for the future.
☐ Not at all   ☐ Well   ☐ Very well

UNIT 12 | I'm going on a cruise.

## E YOUR STORY: Cindy and Ryan are going to Mexico.

**1** Read and listen. What is Cindy's surprise?

**1 | EARLY EVENING**

**Cindy:** Well, you're looking much better.
**Ryan:** Thanks. I'm feeling better. I think I'm just sick of the cold weather.
**Cindy:** I know what you mean. What a month. You and my mother were both sick. Now I can finally relax.

**2**

**Cindy:** So I made a decision. We're going to go on vacation. We can visit Russell in Mexico. I bought the airline tickets!
**Ryan:** You're kidding! What a great idea! When are we going to leave?
**Cindy:** How does Saturday sound?
**Ryan:** Oh, my goodness. Who's going to watch the cafe?

**3**

**Cindy:** Marisol and Jordan. It's all set.
**Ryan:** I don't believe it! Everyone knows but me! How did you keep it a surprise?
**Cindy:** You're easy to keep secrets from. Now, go buy some clothes for sunny Mexico.

**4 | LATER THAT EVENING**

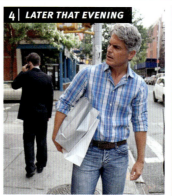

**Peter:** Really?...You're coming to New York?...Married?... Sure, that sounds great... We can talk about it when you're here.
**Ryan:** That's Peter. Who's he talking to?

**5**

**Ryan:** I just saw Peter. He's talking about getting married.
**Cindy:** Well, what do you know! Hmm.... Guess what? I have one more surprise for you!
**Ryan:** We're going to stay at an expensive hotel?
**Cindy:** Even better. My mother is going to come with us.

Online Practice

**2** Listen again. Answer the questions.

1. What did Cindy buy?
2. Who are they going to visit?
3. Who is going to watch the cafe?
4. What does Ryan need to buy?
5. What is Peter talking about?
6. What is Cindy's last surprise?

**3** Use the story to complete the expressions in the box. Listen and check.

**4** Work in a group. Tell your classmates about some surprising news you received.

**Everyday expressions—Showing surprise**

You're _____!
Oh, my _____.
I don't _____ it!
Well, what do you _____!

**> > > > Now I can...** show surprise.
☐ Not at all  ☐ Well  ☐ Very well

## F REAL-WORLD LISTENING: A trip to Honolulu

1. Look at the photo and map. What do you know about Honolulu? What do you think people do there?

2. Complete the sentences about Honolulu. Use the words in the box.

| an island | music | a beach | a city | surfing |

Honolulu, Hawaii is _____ on _____ called Oahu.

Waikiki is _____ in Honolulu with shopping centers and hotels.

Honolulu is a great place for _____ and has traditional _____ and dance.

3. Watch or listen and check your answers.
CD 3-10

4. Watch or listen again. Mark the statements true (**T**) or false (**F**).
CD 3-10
   ____ 1. In the Hawaiian language, Oahu means "gathering place."
   ____ 2. Many people in Honolulu are from other countries.
   ____ 3. In the past, Kamehameha and Waikiki were Hawaiian kings.
   ____ 4. Hawaii became a part of the U.S. in 1908.
   ____ 5. Waikiki has over 650,000 visitors a day.
   ____ 6. The hula is a traditional Hawaiian dance.

5. What are you going to do for your next vacation? Are you going to go to a place like Honolulu? Tell a partner.

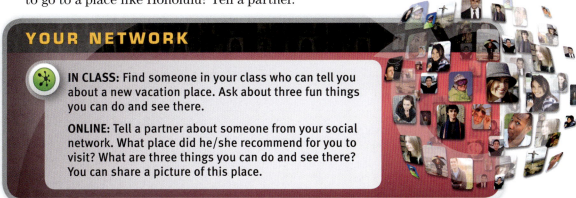

### YOUR NETWORK

**IN CLASS:** Find someone in your class who can tell you about a new vacation place. Ask about three fun things you can do and see there.

**ONLINE:** Tell a partner about someone from your social network. What place did he/she recommend for you to visit? What are three things you can do and see there? You can share a picture of this place.

> > > > **Now I can...** talk about going on a vacation.
☐ Not at all  ☐ Well  ☐ Very well

UNIT 12 | I'm going on a cruise.

# REVIEW Units 9-12

Circle the correct word or words to complete each sentence.

## A | Vocabulary

1. Do you like music? There's a *concert / movie / game* tonight.
2. Is there a *play / game / class* at the theater this weekend?
3. I'm studying French and Spanish. I really like *foreign languages / sciences / arts*.
4. Fatima plays a lot of sports. Her favorite class is *history / physical education / math*.
5. Your *face / neck / nose* is between your head and your chest.
6. I broke my *toe / leg / finger*. Now I can't write.
7. When you travel to another country, you need a *boat / cruise / passport*.
8. Oh, no! It's already 8:15! I hope I don't *travel / miss / lose* the boat!

## B | Grammar

1. Jack *was / were* at home last night.
2. Mike and Linda *wasn't / weren't* in class yesterday.
3. A: Were you in Lima last week?
   B: Yes, I *was / were*.
4. Lisa is a teacher. Last year she *took / taught* art.
5. We went to the park yesterday. We *didn't go / don't went* to the shopping mall.
6. You didn't *like / liked* the movie.
7. *Do / Did* they go skiing yesterday?
8. Where did you *stayed / stay*?
9. What time did you *get up / got up*?
10. Mika *is / are* going to leave soon.
11. Are you going *work / to work* tomorrow?
12. Yes, *I'm going / I am*.

### Now I can...

talk about places to go (page 59)

talk about school subjects (page 65)

name parts of the body (page 71)

talk about travel (page 77)

say where people were (page 60)

talk about the past (page 66)

ask questions about past events (page 72)

talk about my plans for the weekend (page 78)

## C | Reading

 1  Read and listen to Ann's story. Then number the pictures in the correct order.

| Home | News | Business | Sports | Entertainment | Health | Blog | Chat |

### Ann's Trip to England

Last year my sister Rose and I went on a trip to England. First, we went to London. I really liked it. The weather wasn't very nice, but there was a lot to do, so we didn't mind. We saw two plays. We also went to several museums. Rose studied art in college, so she really enjoyed the art museums. I'm taking classes in theater, so I liked the plays the best. We both really liked shopping. There are many places to shop, including Camden Market and Portabello Road. These markets are really crowded on the weekends. They are good places to watch people.

After a week in London, we went to the Lake District in England. It was very beautiful. We stayed at a very small hotel. It had great food. Every day we walked on the paths all over the area. We even visited the home of Beatrix Potter. She wrote *Peter Rabbit* and other books for children. We had a great time, and it was nice to relax after our busy time in London. We had one problem. My sister left her passport at the hotel, so we missed our train back to London.

Next year, we are going to go to Spain. First, we're going to stay in Barcelona for a few days. Rose wants to visit the art museum there, and I want to look at the buildings and parks. It's a very beautiful city. We're also going to relax at the beach nearby. Rose has a friend in Barcelona, so we're going to stay in her apartment. We aren't going to spend money on a hotel in Barcelona, so we'll have enough money to go to Madrid, too. I'm very excited about the trip!

London  Lake District  Barcelona

2  Read the story again. Complete the sentences with the correct word or words.

1. Ann is studying _____ in college.
2. _____ liked the museums in London very much.
3. Both Ann and Rose liked _____.
4. They stayed in London for _____(s).
5. After London, they went to the _____.
6. _____ wrote the book *Peter Rabbit*.
7. Rose left her _____ at the hotel.
8. Next summer, they are going to travel in _____.
9. First, they're going to go to _____.
10. In Barcelona, they are going to stay at _____.

# GET CONNECTED

## Blogging

 The word *blog* is a short form of *weblog*. A blog is a website with information that is posted every day or every week. Usually one person writes a blog, but sometimes many people write for the same blog. Each piece of writing is called a *post*. The posts are in time order. They start with the newest and go back in time.

Blogs usually have:

>> a main area for writing
>> a list of old blog posts
>> a place for people to comment
>> links to other sites

Blogs can be about many things. They can talk about political issues or give news. They can be someone's personal thoughts. Or they might be about specific topics, such as sports, gardening, or healthy foods for children.

**GET Started**

1. Look at the blog on page 85. What is the topic? Do you think it's interesting? Why or why not? Discuss with a partner.

2. Read the blog. Why does the blogger like slides? Write two reasons.

   1. _____

   2. _____

**GET Together**

3. Read the comments by Victor and Jin. Who do you agree with? Why? Discuss with a group.

4. What technology (computers, audio, video, slides, etc.) do you like to use to learn English? What do you think works in a classroom? What works when you study at home? Discuss your ideas with a different group.

# Carly's English Classroom

Home | About | Past blogs | Photos | Links

## Technology in a Language Class

I don't really like to use a lot of technology in my class. I think my students need to speak more, not just play with computers. It's hard for them to turn off their cell phones! But I really like to use slides in my classes now. I heard about them from Annie's blog. Students learn better when they see pictures and hear language at the same time. It is much better than the teacher talking with no pictures. I can also put the slides online. My students can look at them for homework or to review.

### Responses

**Victor** says: This is very interesting. I'm a student, and I learn best with pictures. When my teacher uses slides, I understand a lot. They are a great idea. However, I also like to use computers.

**Jin** says: We don't have a computer in our classroom. The teacher writes on the board. I don't think we need technology. This works for me.

**Leave a comment**
Name:
E-mail:
Comment:

**GET To It!**

**5** Write a comment below in response to the blog. Use your ideas from Activities 3 and 4.

**Leave a comment**
Name:
E-mail:
Comment:

> > > > **Now I can...** understand and comment on blogs.
☐ Not at all   ☐ Well   ☐ Very well

**Take it online**
Read and comment on a blog.

# UNIT 13 I eat a lot of cake.

## YOUR NETWORK

Go to Network Online Practice to record your voice in the conversations on pages 86 and 90.

Go to Network Online Practice to watch video about food trucks in New York City.

Network! Go online to learn about what someone from another country likes to eat. Share on page 91.

Online Practice

### A CONVERSATION: How many cookies do you eat?

1 Look at the picture. What is Ryan doing?

2 Read and listen.
CD 3-12

**Cindy:** What's that?

**Ryan:** It's a questionnaire from the doctor. It's about my diet.

**Cindy:** Uh-huh. "What do you eat and drink in a week?" Number one: How much cheese do you eat?

**Ryan:** Not much.

**Cindy:** Not much? You have a cheese sandwich every day. Number two: How many cookies do you eat?

**Ryan:** Oh, not many.

**Cindy:** Not many? You usually have three or four cookies with a cup of coffee. And…

**Ryan:** You can fill in the questionnaire. I'm going to get some food.

3 Work in pairs. Practice the conversation.

4 Stand and ask your classmates about what they eat.

> How much cheese do you eat?
> Not much./I eat a lot.

> How many cookies do you eat?
> Not many./I eat a lot.

**Now I can...** talk about my diet.
☐ Not at all   ☐ Well   ☐ Very well

86

## B VOCABULARY: Food

**1** Listen and repeat. Then check ✓ the kinds of food you often eat.

☐ 1. beef        ☐ 2. chicken      ☐ 3. fish
☐ 4. cheese      ☐ 5. potatoes     ☐ 6. carrots
☐ 7. apples      ☐ 8. tomatoes     ☐ 9. bread
☐ 10. lettuce    ☐ 11. eggs        ☐ 12. milk

**2** Add more foods and drinks to the chart.

| Food | Drink |
| --- | --- |
| apples | milk |

**3** Talk with a partner. What food and drinks do you like? What don't you like?

**Example:**
A: *Do you like chicken?*
B: *Yes, I do. Do you?*

> > > > > **Now I can...** talk about foods and drinks.
☐ Not at all   ☐ Well   ☐ Very well

UNIT 13 | I eat a lot of cake.      87

## GRAMMAR: Count and noncount nouns

| Count nouns | Noncount nouns |
|---|---|
| Count nouns have a singular and a plural form. | Noncount nouns have only one form. |
| **a** cookie    **some** cookies | **some** cheese |
| We use *many* with count nouns. | We use *much* with noncount nouns. |
| How **many** cookies do you eat? | How **much** cheese do you eat? |
| I don't eat **many** cookies. | I don't eat **much** cheese. |
| We use *a lot of* with both count and noncount nouns. ||
| You eat **a lot of** cookies. ||
| You eat **a lot of** cheese. ||

Grammar Reference page 126

Online Practice

**1** Look at the conversation on page 86. Circle the count nouns. Underline the noncount nouns.

**2** Write *a/an* before the singular count nouns and *some* before the noncount nouns.

1. ___a___ sandwich
2. ___some___ juice
3. _____ water
4. _____ potato
5. _____ money
6. _____ apple
7. _____ snack
8. _____ bread
9. _____ orange
10. _____ milk

**3** Complete the questions with *much* or *many*.

1. How _____ coffee do you drink?
2. How _____ housework do you do?
3. How _____ text messages do you send every day?
4. How _____ cousins do you have?
5. How _____ exercise do you do?
6. How _____ snacks do you eat every day?

**4** Ask a partner the questions from Activity 3.

**Example:**

A: *How much coffee do you drink?*

B: *Not much. / I drink a lot of coffee. / I don't drink any coffee.*

### Pronunciation: *of*

**We usually reduce *of* to /əv/ in expressions. Listen and repeat.**

1. a bowl of rice
2. a lot of potatoes
3. a carton of eggs
4. a cup of coffee
5. a lot of chicken
6. a bag of apples
7. a piece of bread
8. a glass of juice

> > > > **Now I can...** talk about count and noncount nouns.    > > > >
☐ Not at all   ☐ Well   ☐ Very well

## D  READING AND SPEAKING

CD 3-16

**1** Read and listen. Which breakfast does each person eat? Write the correct name under each picture.

# Breakfast around the World

### Tuyen, Vietnam
My breakfast is a bowl of **noodles** or rice with some fish or sometimes meat—usually beef. And at every meal in Vietnam we have a plate of fresh vegetables and **herbs**. A lot of people drink tea or coffee with their breakfast, but I like a cup of **hot chocolate**.

### Connor, Australia
I'm always very **hungry** in the morning, so I like to have a big breakfast. I often have **fried** eggs and **toast** with **jam**. Sometimes I also have fruit—usually apples or **grapes**. I always have coffee and a big glass of orange juice. Breakfast is my favorite meal of the day.

### Mayu, Japan
Like many people in Japan, I usually eat a small breakfast. Because I have a pretty late dinner, I'm usually not very hungry in the morning. I often have bread with jam and butter. Sometimes I have eggs on my bread. I usually drink coffee, but sometimes I have tea.

### Kevin, United States
I don't like to get up early, so I don't have much time for breakfast. I usually have breakfast **on the way** to work. I get a cup of coffee and a **muffin** at a cafe, and I eat it on the train. It's not a very **healthy** breakfast, but it's fast and easy! On weekends, I usually have a big bowl of **cereal** and some **yogurt**.

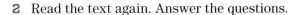

Online Practice

**2** Read the text again. Answer the questions.
1. What do people eat every day in Vietnam?
2. Why does Connor eat a big breakfast?
3. Why does Mayu usually have a small breakfast?
4. Where does Kevin usually eat breakfast?
5. Which breakfast do you think is the healthiest? Why?
6. Which breakfast is most similar to your usual breakfast?

| Language note: Quantities |  |
|---|---|
| With noncount nouns, we describe quantities like this: | |
| rice | a bowl of rice |
| tea | two cups of tea |
| milk | three glasses of milk |
| toast | four pieces of toast |

**3** Work in a group. Talk about your typical breakfast, lunch, and dinner.

*For breakfast, I usually / often have…*  *For dinner, I usually / often have…*

*For lunch, I usually / often have…*  *My favorite meal of the day is…because…*

> > > > **Now I can…** describe what I eat and drink.
☐ Not at all  ☐ Well  ☐ Very well

UNIT 13 | I eat a lot of cake.

# E YOUR STORY: Marisol meets Matt.

**1** Read and listen to the story. What does Marisol think about Matt?

**Jordan:** Hi, Matt. I'm surprised you're up so early.
**Matt:** I was at the gym. Now I'm really hungry. So, what's good here?
**Jordan:** Everything! But the muffins are delicious. Are you ready to order?
**Matt:** Can I have a blueberry muffin?
**Jordan:** Anything to drink?
**Matt:** Coffee, please.

**Jordan:** Here you go. Enjoy.
**Matt:** Thanks. Can I have some eggs, too?
**Jordan:** Um. Sure. I'll put the order in.
**Matt:** And maybe some wheat toast?
**Jordan:** Anything else?
**Matt:** Well, could I also get some fruit and yogurt?

**Marisol:** Here's the fruit and yogurt.
**Jordan:** Thanks, Marisol. It's for my brother over there!
**Marisol:** You didn't tell me you had a brother. He looks nice. I'll take this to him.

**Marisol:** Oh, Matt. That's so interesting! Tell me more!
**Jordan:** Uh, Marisol? I think something is burning in the kitchen.

**2** Listen again and complete the sentences.

Matt was at the _____ before he came to the cafe. Jordan says the blueberry muffins are _____. Matt orders _____ to drink. Matt also wants some wheat _____. _____ is going to bring out his order.

Marisol thinks Matt looks _____.

**3** Use the story to complete the expressions in the box. Listen and check.

**4** Work in a group. Talk about your favorite restaurant meal.

### Everyday expressions—At a restaurant

**Server**

Are you ready _____ order?

(Would you like) anything _____ drink?

**Customer**

Can I _____ a blueberry muffin, please?

Could I also _____ some fruit and yogurt?

> > > > **Now I can...** order a meal in a restaurant.
☐ Not at all   ☐ Well   ☐ Very well

#  REAL-WORLD LISTENING: New York City food trucks

1. Look at the photo of Joe. What is his job?

2. Watch or listen and complete the sentences about Joe. Use the words in the box.

| Belgium | waffles | office buildings |
| New York City | breakfast | a food truck |

Joe works in _____. He sells _____ in _____. Most of his customers work in _____. Waffles are originally from _____, and people eat them for _____ or lunch.

"In New York you can get any sort of food from a food truck."

Online Practice

3. Watch or listen again. Check ✓ the things Joe puts on the waffles.
   - ☐ bananas
   - ☐ dumplings
   - ☐ cheese
   - ☐ Korean food
   - ☐ chicken
   - ☐ strawberries
   - ☐ chocolate
   - ☐ tacos

4. Work in groups. Make a list of street food you can eat in your town. How often do you eat these things?

**YOUR NETWORK**

**IN CLASS:** Interview three classmates. What are three special foods each person likes to eat? What is one special dish he/she can cook?

**ONLINE:** Tell a partner about someone from your social network. What country is he/she from? What kind of food from his/her country does he/she like to eat? You can share a picture or a recipe of that food.

> > > > **Now I can...** talk about the kinds of food I can eat in my town.
☐ Not at all  ☐ Well  ☐ Very well

# UNIT 14 What do you like to wear?

## YOUR NETWORK

Go to Network Online Practice to record your voice in the conversations on pages 92 and 96.

Go to Network Online Practice to watch video about fashion and fabrics.

Network! Go online to ask someone what he/she likes to wear on the weekend. Share on page 97.

Online Practice

### A CONVERSATION: What are you going to wear tonight?

1 Look at the picture. What are Ryan and Cindy doing?

2 Read and listen.
CD 3-20

**Cindy:** What are you going to wear tonight to the concert?
**Ryan:** I'm not sure.
**Cindy:** At least you're not wearing that old brown jacket.
**Ryan:** I like that jacket. It's comfortable. But where is it?
**Cindy:** Oh, why don't you wear your new blue jacket?
**Ryan:** I don't want to wear the old brown jacket, but I left the concert tickets in one of the pockets.
**Cindy:** Oh, no. I gave it away!

3 Work in pairs. Practice the conversation.

4 Work in pairs. Practice the conversation again. This time, replace "concert" and "old brown jacket."

> What are you going to wear tonight to the _____?

> I'm not sure.

> At least you're not wearing that _____.

> I like that jacket.

**Now I can...** describe clothing.
☐ Not at all ☐ Well ☐ Very well

# B VOCABULARY: Clothes

1 Listen and repeat. Then check ✓ the clothes you are wearing now.

☐ 1. a belt
☐ 2. socks
☐ 3. a hat
☐ 4. shorts
☐ 5. a shirt
☐ 6. a jacket
☐ 7. pants
☐ 8. shoes
☐ 9. a blouse
☐ 10. jeans
☐ 11. boots
☐ 12. a dress
13. tights
14. a skirt
15. a coat
16. gloves

Online Practice

2 Think of more words for clothes. Share them with a partner.

3 Listen. Write questions about the price. Use *How much…*

**Examples:**
You hear: *Do you like this jacket?*
You write: *Yes. How much is it?*
You hear: *Do you like these shorts?*
You write: *Yes. How much are they?*

> **Language note: Plural words**
> These words are always plural:
> pants   jeans   shorts   tights
> How much **is this** shirt? **It's** $20.
> How much **are these** pants? **They're** $30.

### Pronunciation: /s/ and /z/

1. **Listen and repeat.**

　　　/s/　　　/z/
　　　dress　　shoes

2. **Write /s/ or /z/ next to each word.**

　___ this　　___ glass　　___ shorts　　___ music　　___ easy
　___ these　___ skirt　　___ cousin　　___ sock　　___ please

3. **Listen and repeat. Were your answers to Part 2 correct?**

> > > > > Now I can... talk about clothes.
☐ Not at all　☐ Well　☐ Very well

UNIT 14 | What do you like to wear?

## C GRAMMAR: Adjectives

**Grammar Reference page 126**

| Adjectives | |
|---|---|
| **Adjectives go before a noun.** | **Adjectives are the same for singular and plural nouns.** |
| a **brown** jacket NOT a jacket brown | an **old** shirt   **old** shirts   NOT olds shirts |
| **new** clothes NOT clothes new | |

| Adjective order | | | | Examples |
|---|---|---|---|---|
| **1. quality** | **2. size** | **3. age** | **4. color** | that **old brown** jacket |
| good | big | old | black | a **nice long** vacation |
| bad | small | new | white | those **big old** houses |
| beautiful | long | | blue | a **beautiful red** dress |
| nice | short | | red | |

Online Practice

**1** Look at the conversation on page 92. Circle the adjectives.

**2** Choose the correct adjective order.

1. He's a _____ actor.
   a. *young handsome*      b. *handsome young*

2. I am going to wear my _____ dress.
   a. *long black*      b. *black long*

3. She wants a _____ hat.
   a. *green beautiful*      b. *beautiful green*

4. He cooked a _____ dinner for us.
   a. *Mexican wonderful*      b. *wonderful Mexican*

**3** Complete the sentences. Put the words in parentheses in the correct order.

1. She has ___*long black hair*___ *black / hair / long*.
2. He drives a _____ *car / old / dirty*.
3. This is my _____ *nice / office / new*.
4. Can you give me that _____ *blue / small / book*, please?
5. How much are these _____ *purses / nice / red*?
6. The baby has _____ *eyes / brown / beautiful*.

**4** Listen. Were your answers to Activity 3 correct?

**5** Complete the sentences about yourself. Use two adjectives in each sentence.

1. Right now, I'm wearing _____.
2. In my bag/pockets, I have _____.
3. I have _____ hair.
4. I live in a(n) _____ apartment/house.

**6** Work in a group. Share your sentences from Activity 5.

>>>>> **Now I can...** describe things.
☐ Not at all   ☐ Well   ☐ Very well

94

# D  READING AND SPEAKING

1  Read and listen. Write the correct name after each description.

## Work Clothes

**A:** I work at a nightclub. When I'm at home, I wear casual clothes—usually jeans and a T-shirt. When I go to work, I wear a nice black suit, a white shirt, a black tie, and black shoes. I stand outside the club, so I sometimes wear a coat and black gloves, too.

**B:** I work for an airline. I wear a uniform to work. Our uniform is nice and colorful. We wear green pants and a yellow shirt with a green and yellow scarf. We have a jacket, too, but I don't wear it all the time. I also wear a badge with my name on it.

**C:** I work at a gym. For work, I wear a white T-shirt and red shorts with socks and sneakers. When I'm not at work, I like wearing cool clothes—usually pants and a shirt with a tie.

**D:** I'm a fashion model, so I wear lots of different clothes every day—dresses, skirts, blouses, shoes…This week, I'm doing some work for a shoe company. Right now, the photographer is taking photographs of my long purple boots. Aren't they gorgeous?

Imelda

Katrina

Imran

Carlos

Online Practice

2  Answer the questions.

|  | Imelda | Katrina | Imran | Carlos |
|---|---|---|---|---|
| 1. What does each person do? |  |  |  |  |
| 2. What does each person wear to work? |  |  |  |  |
| 3. What does each person think about his/her work clothes? |  |  |  |  |

 3  Tell a partner about your clothes.

*Yesterday, I wore…*

*When I go to work/school, I wear…*

*On weekends, I usually wear…*

*When I was in elementary school, I wore…*

> > > > **Now I can…** describe people's clothes.
☐ Not at all  ☐ Well  ☐ Very well

UNIT 14 | What do you like to wear?

## E YOUR STORY: Lucy gets a new job.

**1** Read and listen to the story. Where is Lucy going?

**1 | IN THE MORNING**

**Peter:** Hello, Lucy. You look very nice today.
**Lucy:** Oh, hello, Peter. Thank you. I have an interview for a new job.
**Peter:** How exciting! What's the job?
**Lucy:** Assistant copywriter for an advertising agency.
**Peter:** What do copywriters do?
**Lucy:** They help write ads.
**Peter:** That's great. Do you have time for coffee?
**Lucy:** Yes, I have a few minutes.

**2**

**Lucy:** So, why are you here this morning? You're usually at work by now.
**Peter:** I'm meeting someone here at 10:00.
**Lucy:** Anyone I know?
**Peter:** No, well, actually it's my ex-girlfriend, Amanda. She wants to talk about plans for the future.
**Lucy:** Hmm. That sounds interesting. Where's Sarah?
**Peter:** She's out of town.

**3 | LATER THAT DAY**

**Jordan:** How was the interview?
**Lucy:** Great! I got the job.
**Jordan:** Hurray! Let's celebrate. Do you want to call Peter and Sarah?
**Lucy:** I think Sarah's out of town and Peter's with his ex-girlfriend.
**Jordan:** Oh, really?

**Peter:** Yes, Amanda. It was good to see you, too. And again, congratulations. I hope you have a great wedding. Thanks for returning my CDs.

Online Practice

**2** Listen again. Read the statements. Write true (**T**) or false (**F**).

____ 1. Lucy has an interview with a marketing company.

____ 2. Peter has a meeting at 10:00 at the cafe.

____ 3. Peter is meeting his sister, Amanda.

____ 4. Sarah and Peter are out of town.

**3** Use the story to complete the expressions in the box. Listen and check.

**4** Work in a group. Lucy is excited about her new job. Talk about a time when you were excited about something. Make comments as you hear your classmates' stories.

**Everyday expressions—Making comments**

You look very _____.
How _____!
That's _____.
That _____ interesting.
Oh, _____?

> > > > > **Now I can...** make comments.
☐ Not at all  ☐ Well  ☐ Very well

## REAL-WORLD LISTENING: Fashion and fabrics

1. Look at the photos. Do you like the clothes? When do you get dressed up?

2. Complete the sentences with the words in the box.

| sheep | traditional | fabrics | silkworm | plants | patterns |

People make clothing from silk, wool, and other _____. Some fabrics come from _____ like cotton, and other fabrics come from animals like sheep. One _____ can make up to three miles of silk, and one _____ can make wool for a large sweater. In India and in other places, people still use _____ ways to make fabric. The _____ on these fabrics often have special meanings.

3. Watch or listen and check your answers.
   CD 3-30

4. Watch or listen again. Check ✓ the sentences that the reporter in the video probably agrees with.
   CD 3-30
   - ☐ 1. Traditional fabrics can never be trendy.
   - ☐ 2. People in Polynesia make a good useful fabric from flax.
   - ☐ 3. The average person has many pieces of cotton clothing.
   - ☐ 4. Silk looks beautiful and expensive.
   - ☐ 5. Patterns don't have special meanings in the U.S. or the UK.
   - ☐ 6. People can't design good patterns with computers.

5. What fabrics are you wearing now? Do you know where they come from? Tell a partner.

### YOUR NETWORK

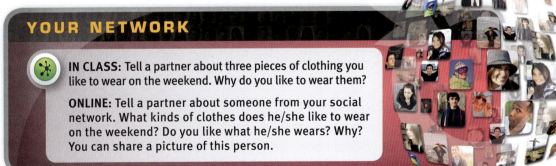

**IN CLASS:** Tell a partner about three pieces of clothing you like to wear on the weekend. Why do you like to wear them?

**ONLINE:** Tell a partner about someone from your social network. What kinds of clothes does he/she like to wear on the weekend? Do you like what he/she wears? Why? You can share a picture of this person.

> > > > > **Now I can...** talk about fabrics and clothes.
> ☐ Not at all   ☐ Well   ☐ Very well

# UNIT 15 My hometown is nicer.

## YOUR NETWORK

Go to Network Online Practice to record your voice in the conversations on pages 98 and 102.

Go to Network Online Practice to watch video about interesting places in Chile.

Network! Go online to find someone who lives in a very cold or very hot place. Share on page 103.

Online Practice

### A CONVERSATION: How was your weekend?

1. Look at the picture. What are Yuka and Sarah doing?

2. Read and listen.
   CD 3-31
   **Yuka:** How was your weekend in Washington, D.C.?
   **Sarah:** Well, it was fun, but the weather wasn't very good.
   **Yuka:** Too bad. It was warm here with only a few clouds.
   **Sarah:** I heard. It was cloudier in Washington, D.C. than here—and colder, too.
   **Yuka:** Yes. And now they say it's going to rain tomorrow.
   **Sarah:** Just my luck!

3. Work in pairs. Practice the conversation.

4. Stand and tell your classmates about the weather last weekend.

   How was your weekend?  How was the weather?
   It was fun./Not very good.  It was warm/cold.

**Now I can...** talk about the weather.
☐ Not at all  ☐ Well  ☐ Very well

**VOCABULARY:** The weather

1 Listen and repeat. What is the weather like today?

1. sunny

2. rainy

3. cloudy

4. windy

5. snowy

6. foggy

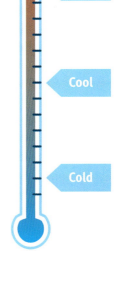

2 Think of more weather words. Share your ideas with a partner.

3 Talk with a partner. What's the weather like in your favorite place?
- today
- in January
- in April
- in June
- in August
- in October

**Language note: Word building**

| Noun | Adjective | Noun | Verb |
|---|---|---|---|
| clouds | cloudy | rain | to rain |
| wind | windy | snow | to snow |
| a storm | stormy | | |

**Pronunciation: Vowel sounds**

1. Check ☑ the pairs with the same vowel sound.
   - ☐ 1. warm   storm
   - ☐ 2. big   nice
   - ☐ 3. short   cold
   - ☐ 4. hot   worse
   - ☐ 5. winter   windy
   - ☐ 6. cool   good

 2. Listen and repeat. Were your answers to Part 1 correct?

> > > > > **Now I can...** talk about the weather in my favorite place.
☐ Not at all    ☐ Well    ☐ Very well

## C GRAMMAR: Comparatives

### Grammar Reference page 127

| Comparatives | | | |
|---|---|---|---|
| Yesterday was **warmer than** today. | | My new job is **more interesting than** my old job. | |
| Base | Comparative | Base | Comparative |
| warm | **warmer** | interesting | **more interesting** |
| nice | **nicer** | windy | **windier** |
| hot | **hotter** | | |
| **Irregular** | | | |
| good    **better** | | bad    **worse** | |

Online Practice

**1** Look at the conversation on page 98. Circle the comparatives.

**2** Write the comparatives.
1. It was cold yesterday.                It's ___colder___ today.
2. It was cloudy yesterday.              It's _____ today.
3. It was hot yesterday.                 It's _____ today.
4. The weather was good yesterday.       It's _____ today.
5. The weather was bad yesterday.        It's _____ today.
6. The roads were dangerous yesterday.   They're _____ today.

**3** Write two comparative sentences for each item. Use the adjectives in parentheses.
1. Chris is 22 years old. Mike is 25 years old.
   _Chris is younger than Mike. Mike is older than Chris._ (young/old)
2. It's 35 degrees in Taipei. It's 30 degrees in Beijing.
   _____ (hot/cool)
3. It's rainy and cold in Boston. It's sunny and warm in Miami.
   _____ (good/bad)
4. The brown jacket is $50. The red jacket is $75.
   _____ (cheap/expensive)

**4** Work with a partner. Make two comparisons for each pair. Use ten different adjectives.

**Example:**
A: *Shopper's World is bigger than Wilson's.*
B: *Wilson's is more expensive than Shopper's World.*

- two stores in your town
- two famous people
- two people in your family
- two restaurants in your town
- two TV shows
- two cities

>>>>> **Now I can...** compare two things.
☐ Not at all    ☐ Well    ☐ Very well

# D   READING AND WRITING

1  Read and listen. Where is Natsuki from? Where does she live now? Does she like her new home?

## SEATTLE, My New Home

Natsuki is from Osaka, Japan. Last year, she moved to the United States with her husband and two children. They now live in Seattle, on the northwest coast. Natsuki works for an **electronics** company there.

"When I tell people that I live on the west coast of the U.S., they always think of California, and they think it's warm and sunny here. But that's **further** south. It isn't like that in Seattle. In fact, the weather in Osaka is better. The winters in Osaka are about the same as here, but the summers there are hotter and drier. Here in Seattle, it's usually gray and cloudy, and it rains a lot.

My **salary** in Osaka was higher, but my job here is more interesting. I have a shorter work day here, so I can spend more time with my family. Also, cars, gas, and houses are all cheaper in the U.S. In Osaka, we only had a small apartment, but here we have a big house with a **yard**. Of course, everything is bigger in the States—the houses, the cars, the stores, the meals!

However, some important things are more expensive. Medicine is very expensive, and you have to pay for **health insurance**, too.

We miss our family, but they come and visit us. What does my husband miss most? Soccer. It isn't a big sport in the States. They play **American football**, baseball, and basketball here. But we love life in Seattle. We have a lot of good friends."

Online Practice

2  Read the statements. Write (**S**) for Seattle or (**O**) for Osaka.

_O_  1. The weather is warmer.  
____  2. She earns less money.  
____  3. Cars and gas are more expensive.  
____  4. Houses are bigger.  
____  5. Medicine is cheaper.  
____  6. Soccer is more popular.

3  Write a comparison between two places. Use an idea from below.

- two places you go on vacation
- two parts of your country
- two countries
- two cities

**Example:**

*Portland is better than San Diego because it's more interesting.*

*Portland is colder than San Diego, but San Diego is sunnier.*

*Restaurants in San Diego are cheaper than in Portland, but they are better in Portland.*

**Language note: Compass directions**

4  Share your comparison with a partner.

> > > > > **Now I can...** compare different places.
☐ Not at all   ☐ Well   ☐ Very well

## E YOUR STORY: Ryan and Cindy come home from Mexico.

**1** Read and listen to the story. How was Ryan and Cindy's trip?

**Jordan:** Hi, Ryan. Welcome home. When did you get back?
**Ryan:** Late last night.
**Jordan:** Are you and Cindy tired?
**Ryan:** We're OK, but it's a lot colder here than in Mexico.
**Jordan:** How did it go?
**Ryan:** Well, the weather was terrific, but we didn't see much of Russell. He started playing in a band there.

**Ryan:** How was business?
**Jordan:** Good. We were busy, but we didn't have any problems.
**Ryan:** Anything interesting happen while we were away?
**Jordan:** Hmm. Let's see. My brother is still here, unfortunately. Sarah went to Washington, D.C. for the weekend.

**Ryan:** Hi, Lucy.
**Lucy:** Hi, Ryan. Welcome back.
**Ryan:** Thanks. I heard you have a new job. Do you like it?
**Lucy:** Yes. It's so interesting. I also like my new boss. She's much nicer than my old boss. Her name is Rose Brown.

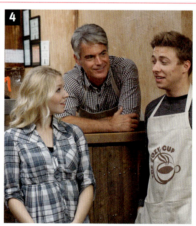

**Ryan:** That's great. Anything else going on?
**Jordan:** Well, I think Peter's old girlfriend is in town.

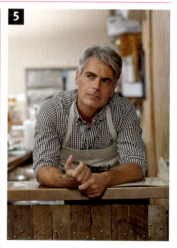

**Ryan:** Really? So that's who Peter was talking to on the phone!

**2** Listen again and complete the sentences.

1. The weather in Mexico was _____.
2. Russell is playing _____ in Mexico.
3. The cafe was _____ while they were away.
4. Lucy's new job is _____.
5. Lucy's new boss is _____ than her old boss.

**3** Use the story to complete the expressions in the box. Listen and check.

**4** Work in a group. Talk about a trip you took. How was it? How was the weather?

---
**Everyday expressions**—Talking about a trip

Welcome _____.
When did you _____ back?
Are you _____?
_____ did it go?

---

**>>>> Now I can...** talk about a trip.
☐ Not at all  ☐ Well  ☐ Very well

# F  REAL-WORLD LISTENING: Around the world—Chile

1. Look at the photo and map. What do you know about Chile? What is the weather like there?

2. Watch or listen to a report about places in Chile. Where can you see these things? Write **S** (Santiago), **A** (Atacama), or **P** (Patagonia).

   ____ 1. the desert      ____ 3. the Valley of the Moon      ____ 5. Kilometer Zero
   ____ 2. skyscrapers     ____ 4. Gray Lake Glacier           ____ 6. horses

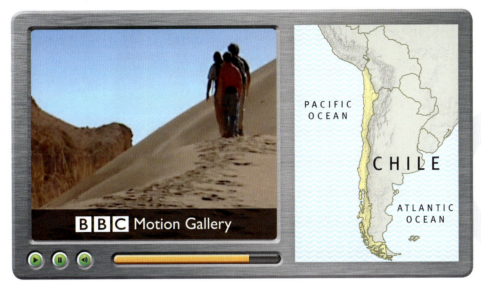

1 mile = 1.6 kilometers

3. Watch or listen again. Complete the sentences. Use the words in the box.

   | bigger | drier | smaller | more crowded | more beautiful | more unpredictable |

   1. Santiago is _____ than the Atacama Desert.
   2. The Atacama Desert is _____ than Santiago and Patagonia.
   3. In the Valley of the Moon, the views are _____ at sunset.
   4. Patagonia is _____ than the Atacama Desert.
   5. The weather in Patagonia is _____ than the weather in Atacama.
   6. The Gray Lake Glacier gets _____ every year.

4. Which place do you prefer, Santiago, Patagonia, or the Atacama Desert? Why? Tell a partner. Use comparisons in your answer.

   **Example:**
   *I prefer the Atacama Desert because it's drier than the other places. I hate rain!*

### YOUR NETWORK

**IN CLASS:** Tell a partner about the kind of weather you like and why. What are three activities you like to do in that weather?

**ONLINE:** Tell a partner about someone from your social network. Where is the person from? How cold or hot does it get in his/her town or city? You can share a picture of this place.

> > > > > **Now I can...** talk about preferences.
☐ Not at all   ☐ Well   ☐ Very well

UNIT 15 | My hometown is nicer.   103

# UNIT 16 Around the world

## YOUR NETWORK

Go to Network Online Practice to record your voice in the conversations on pages 104 and 108.

Go to Network Online Practice to watch video about famous places in New York City.

Network! Go online to find someone who lives near a famous place. Share on page 109.

Online Practice

### A CONVERSATION: The Amazon and Copacabana

1. Look at the picture. What do you think they are doing?

2. Read and listen. *CD 3-39*

   **Matt:** What are you doing?
   **Sarah:** Some research for a presentation on Brazil.
   **Matt:** Well, at least you know a lot about it already.
   **Sarah:** True. Do you know which river is the widest in the world?
   **Matt:** The Amazon?
   **Sarah:** Right! And Rio de Janeiro has the most famous beach in the world—Copacabana.

3. Work in pairs. Practice the conversation.

4. Write three questions about interesting places. Then stand and ask your classmates your questions.

> Do you know which river is the widest in the world?
>
> No, I don't./Yes, I do. It's _____.

> Do you know which beach is the most famous?
>
> No, I don't./Yes, it's _____.

**Now I can...** talk about interesting places.
☐ Not at all  ☐ Well  ☐ Very well

## B VOCABULARY: Geographical features

1 Listen and repeat. Check ✓ the features your town or city has.

- ☐ 1. the ocean
- ☐ 2. a mountain
- ☐ 3. an island
- ☐ 4. a cliff
- ☐ 5. a hill
- ☐ 6. railroad tracks
- ☐ 7. a lake
- ☐ 8. a forest
- ☐ 9. a bridge
- ☐ 10. a tunnel
- ☐ 11. a valley
- ☐ 12. a road
- ☐ 13. a river
- ☐ 14. a field
- ☐ 15. a highway

**Online Practice**

2 Write the words from Activity 1 in the correct column.

| Natural features | Man-made features |
| --- | --- |
| the ocean | railroad tracks |

3 Write five sentences about your town. Use the words in the picture.

**Examples:**
1. *Our town isn't near the ocean.*
2. *There's a river in our town. Its name is the Orinoco River.*

> > > > > **Now I can...** talk about geographical features.
☐ Not at all  ☐ Well  ☐ Very well

UNIT 16 | Around the world 105

## C GRAMMAR: Superlatives

Grammar Reference page 127

### Superlatives

We use the superlative form of adjectives to compare more than two things.

It's the **highest** mountain in the world.

| Base | Comparative | Superlative |
|---|---|---|
| high | higher | the highest |
| wide | wider | the widest |
| wet | wetter | the wettest |
| dry | drier | the driest |
| beautiful | more beautiful | the most beautiful |
| **Irregular** | | |
| good    better    **the best** | bad    worse    **the worst** | |

Online Practice

1. Look at the conversation on page 104. Circle the superlatives.

2. Complete the sentences. Use superlatives.
    1. The Pacific Ocean is ___the biggest___ ocean in the world. (big)
    2. Lake Baikal in Russia is _____ lake in the world. (deep)
    3. The Atacama Desert in Chile is _____ desert in the world. (dry)
    4. Mount Everest is _____ mountain in the world. (high)
    5. The Nile River is _____ river in the world. (long)
    6. Mumbai is _____ city in the world. (crowded)

 3. Work with a partner. Look at the chart below. Ask and answer questions about some of the places in Activity 2.

**Example:**
A: The <u>Pacific Ocean</u> is the <u>biggest ocean</u> in the world.
B: Really? How <u>big</u> is it?
A: It's <u>165.2 square kilometers</u>.

| Place | Information |
|---|---|
| Pacific Ocean | 63.8 million square miles (165.2 square kilometers) |
| Lake Baikal | 5,387 feet (1,642 meters) deep |
| Mount Everest | 29,029 feet (8,848 meters) high |
| The Nile River | 4,130 miles (6,650 kilometers) long |

>>>>> **Now I can...** compare more than two things.
☐ Not at all  ☐ Well  ☐ Very well

## D READING AND WRITING

**1** Read and listen. What are the three best places to visit in New Zealand?

### New Zealand
### Come to the Most Beautiful Place in the World!

Are you looking for an exciting vacation? Then come to New Zealand! It has something for everyone—cities with great restaurants and shopping, extreme sports, and beautiful scenery. And if you like peaceful places, New Zealand is for you: with only 4 million people, New Zealand is one of the world's least crowded countries.

**Best places to visit:**

1. Milford Sound is New Zealand's most famous tourist destination. With a yearly rainfall of 6,813 mm (22 feet), it is the wettest place in New Zealand, and one of the wettest places in the world! It has many waterfalls, some a thousand meters (.62 miles) long. Take a boat ride for a beautiful view of the Sound!
2. Mount Cook: At 3,754 meters (4,105 yards), Mount Cook is the highest mountain in New Zealand. In Mount Cook National Park, you can hike, mountain bike, or take a scenic airplane ride.
3. Queenstown: If you like adventure, you will love Queenstown! This lovely seaside town is the home of bungee jumping and jet boating. You can also enjoy Queenstown's many cafes, restaurants, and stores.

**Best time to go:**

Any time is great for a visit to New Zealand. The warmest months are December, January, and February, with average high temperatures between 20º–30ºC (68º–86ºF). The coldest months are June, July, and August, with average high temperatures between 10º–15ºC (50º–59ºF).

Online Practice

**2** Match each topic in the box with a paragraph below.

| Cities | The weather | ~~General information~~ |

**Language note: Paragraph planning**
Group your ideas on a topic together to make a paragraph. When you plan your writing, give each paragraph a topic.

1. Topic: _General information_

   _____ is a (large/small) country with a population of _____ (number).

   The highest mountain is _____ and the longest river is _____.

2. Topic: _____

   The capital city is _____. Here, you can _____ (things to do and see).

   It (is also/isn't) the largest city.

3. Topic: _____

   The weather in _____ is usually _____. In summer, it's _____ and in winter it's _____. The (north/south/east/west) is (warmer/cooler) than the _____.

**3** Complete the three paragraphs above about your country.

> > > > > **Now I can...** describe a country.
☐ Not at all  ☐ Well  ☐ Very well

**YOUR STORY:** Sarah goes to Brazil.

 1  Read and listen to the story. Where is Sarah going?

**1**

**Sarah:** I just got a phone call from home, and I have to go to Brazil.
**Yuka:** What's the problem?
**Sarah:** My sister broke her arm. She needs surgery, so I'm going back for a few days.
**Yuka:** Oh, no! Is she going to be OK?
**Sarah:** Yes, but she wants her big sister there.

**2**

**Sarah:** Hi, Ryan. Is that my coffee and muffin to go?
**Ryan:** Here you go. You look like you're in a hurry.
**Sarah:** Yes, I'm on my way to the airport. I have a flight to Brazil in a couple of hours.
**Ryan:** But…
**Sarah:** Sorry, I can't talk now. My taxi's outside.

**3**

**Ryan:** Hi, Peter. Did you hear the news?
**Peter:** What news?
**Ryan:** The news about Sarah.
**Peter:** What do you mean?
**Ryan:** She's going back to Brazil today.
**Peter:** Brazil? Why?
**Ryan:** Maybe she heard the news about you.
**Peter:** About me? I don't understand.
**Ryan:** Yes. You're going to marry your ex-girlfriend, right?
**Peter:** Amanda? What are you talking about? I'm not marrying anyone.
**Ryan:** But I heard you talking to her about it. She came here to see you.
**Peter:** She just wanted to return some CDs. Amanda's getting married, but to someone named Manuel—not to me.
**Ryan:** Oh, but I thought…

**4**

**Peter:** Taxi! I need to get to the airport. And quickly, please!

Online Practice

 2  Listen again and answer the questions.

1. What happened to Sarah's sister?
2. What does Sarah's sister want?
3. What does Sarah decide to do?
4. Why does Sarah go to the cafe?
5. How is Sarah going to the airport?
6. When does her flight leave?
7. Why does Ryan think Sarah is leaving?
8. What is Peter going to do?

 3  Use the story to complete the expressions in the box. Listen and check.

 4  Work in a group. Discuss these questions.

1. Why is Peter going to the airport?
2. What do you think is going to happen next?

 5  Listen. Were you correct about what happens next?

**Everyday Expressions—Asking for an explanation**

What's the _____?
What do you _____?
Why?
What are you _____ about?

>>>>> **Now I can…** ask for an explanation.
☐ Not at all   ☐ Well   ☐ Very well

##  REAL-WORLD LISTENING: New York City landmarks

1. Look at the photo of Josh. What do you think his job is?

2. Watch or listen to Josh. Does he like his job? Why or why not?

3. Watch or listen again. Complete the sentences. Use the words from the box.

| greatest | largest | most crowded |
| most beautiful | oldest | tallest |

1. The Empire State Building is the _____ building in New York City.
2. Times Square is the _____ part of New York City.
3. Macy's is the world's _____ store.
4. The Brooklyn Bridge is the _____ bridge in New York City.
5. Josh thinks the Flatiron Building is the _____ building in the city.
6. He thinks New York is the _____ city in the world.

4. Work in groups. Make a list of interesting places in your city. Use superlative adjectives to describe them.

*Torre Mayor is the tallest building in Mexico City.*

### YOUR NETWORK

**IN CLASS:** Tell a partner about a place you like to visit. Describe the place. What can you see and do there? Share a picture of this place.

**ONLINE:** Tell a partner about someone from your social network. Where is the person from? What famous place does he/she live near? Describe the place. Share a picture of this place.

>>>>> **Now I can...** describe places to visit in my city.
☐ Not at all  ☐ Well  ☐ Very well

# REVIEW Units 13-16

Circle the correct word or words to complete each sentence.

## A | Vocabulary

1. *Bread / Beef / Milk* is a kind of meat.
2. I like to drink *cheese / tomatoes / orange juice*.
3. It's cold today. I'm going to wear *a jacket / shorts / a belt*.
4. You wear *boots / tights / gloves* on your hands.
5. It's not sunny today. It's *rainy / warm*.
6. We're going skiing. It's very *foggy / snowy* today.
7. A *lake / tunnel / river* is a man-made feature.
8. Australia is a very large *island / ocean / lake*.

**Now I can...**

talk about foods and drinks (page 87)

talk about clothes (page 93)

talk about the weather in my favorite place (page 99)

talk about geographical features (page 105)

## B | Grammar

1. Laura eats *a / an / some* apple every day for lunch.
2. How *much / many* tomatoes do we need?
3. I don't eat *much / many / some* cookies.
4. These pants are *old / olds*.
5. Vera has a *small blue / blue small* car.
6. Jacob has *black long / long black* hair.
7. Today is *hotter than / more hot than / hot than* yesterday.
8. Costa Rica is *more rain / rainier / rainiest* than Syria.
9. The weather in Italy is *more good / better / more better* than in Antarctica.
10. I think Paris is the *most beautiful / the beautifullest* city in the world.
11. The Mississippi is *the most large / largest* river in the United States.
12. The mountains are *the goodest / the most good / the best* place to take a vacation.

talk about count and noncount nouns (page 88)

describe things (page 94)

compare two things (page 100)

compare more than two things (page 106)

## C | Reading

1 Read and listen to Jeff's travel blog. For each city, underline the description of the photograph.

CD 3-47

| Home | News | Business | Sports | Entertainment | **Jeff's Travel Blog** | Chat |

### 1. Hong Kong SAR, China
Hong Kong is a **fascinating** city. It is a very old city, dating back more than 5000 years. But it is also a very modern city. According to a recent survey, it is one of the most important fashion centers in the world. It is a fashion leader
5  in Asia, with **traditional** Chinese clothing as well as the latest styles. But fashion isn't the only thing Hong Kong is famous for. The most popular tourist attraction is The Peak. This is the **highest** mountain on Hong Kong Island. Tourists also like to take a boat around Victoria Harbor or eat food at one of its 11,000 restaurants. When I visited, I really enjoyed the excitement of the city.

10 ### 2. Cairo, Egypt
Cairo is the largest city in Africa and one of the most amazing. It is very close to the Great Sphinx and the Pyramids of Giza. Tourists often visit these **historic** places to learn about the **ancient** civilization. Cairo is on the Nile River, one of the longest rivers in the world. Cairo gets very **hot** in the summer, so the
15 best time to visit is during the **cooler** months between November and March. Cairo's Egyptian Museum has many interesting things to see. I didn't know much about Cairo before I went, and I was very happy to learn some history of the country. I also really liked shopping in the bazaars, or markets.

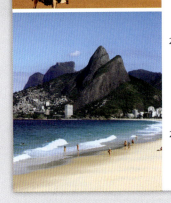

### 3. Rio de Janeiro, Brazil
20 Rio is another one of my favorite cities. It is the second largest in Brazil, but maybe the most famous. People love to go to Rio for the festivals, such as Carnival, its music and dancing, and its beaches. One of the most **famous** beaches in the world is Ipanema Beach. The beach is divided into different sections that are like little neighborhoods. Rio also has beautiful botanical
25 gardens. I'm going to see the Olympics in Rio in 2016.

2 Find the adjectives in bold in the reading. Match the adjective to the noun it describes.

| Adjective | Noun |
|---|---|
| ___ 1. fascinating | a. beaches |
| ___ 2. traditional | b. city of Hong Kong |
| ___ 3. highest | c. civilization |
| ___ 4. historic | d. clothing |
| ___ 5. ancient | e. months |
| ___ 6. hot | f. mountain |
| ___ 7. cooler | g. places |
| ___ 8. famous | h. Cairo |

# GET CONNECTED

## Connecting online

People use social media to meet others from different places. When people go online, they can talk to someone on the other side of the world. But people also use social media to connect with others in their own communities. Sometimes they connect with old friends. Sometimes they meet new people.

On many social networks, you can post your job, your interests, and your city. This helps you find old friends and make new ones. For example, an English student in Rio can find other English students in Rio. A fan of Lady Gaga in London can meet other people in London with the same interest. Some online groups, such as Meetup, help people in one city get to know others with the same interest.

**GET Started**

**1** Look at the Web page for a Meetup group on page 113. What kind of information can you find? Make a list with a partner.

**2** Read the information on the website. Answer the questions.
   1. What city is the group in? _____
   2. Who is the group for? _____
   3. How many members does it have? _____
   4. What kinds of events do they have? _____
   5. How can you sign up for a specific event? _____

**GET Together**

**3** What things do you like to do in your town or city? List three things in the chart below. Then find at least one classmate who has the same interests as you.

| My interest | Classmate's name |
|---|---|
|  |  |
|  |  |
|  |  |

# Come explore Ottawa!

**HOME** | MEMBERS | SPONSORS | PHOTOS | PAGES | MORE | JOIN US!

### OTTAWA, CANADA
This group is for anyone in Ottawa, Canada, who wants to meet new people, make new friends, learn about cultures, and see Ottawa and nearby attractions.

## EVENTS

 **Ottawa River Bike Ride**

Join us as we ride bicycles by the beautiful Ottawa River. There are many birds and flowers to see. Then lunch at a nearby restaurant.

Sun Aug 28 9:00 a.m.
RSVP
5 attending
15 spots left

 **English Club—Advanced Hamden Library**

The English Club is a weekly event. You can improve your English and learn about Canadian culture.

Tue Aug 30 7:00 p.m.
RSVP
4 attending
21 spots left

 **ByWard Market**

Come with us to ByWard Market. It is a great farmer's market and shopping area. There are excellent restaurants here.

Sat Sept 3 10:00 a.m.
RSVP
7 attending
5 spots left

---

Members 2134
Group reviews 121
Upcoming meetups 10
Past meetups 247
What's new?
New member Lidia Martinez

---

**GET To It!**

**4** Work with a partner. Write links for other websites, social networks, etc. where you can meet people with the same interests as you.

> > > > **Now I can...** connect online with people near me.
☐ Not at all  ☐ Well  ☐ Very well

**Take it online**
Find a meetup group near you.

# Audio Scripts

## Unit 1 >>>>>

### VOCABULARY p. 3

**1.**
**Hiroko:** Hi. I'm Hiroko. I'm from Japan.
**Pedro:** Nice to meet you, Hiroko. My name's Pedro. I'm from Mexico.
**Hiroko:** Nice to meet you, too, Pedro.

**2.**
**Ava:** Hello. My name's Ava.
**José:** Nice to meet you, Ava. I'm José. Where are you from?
**Ava:** I'm from the United States. And you?
**José:** I'm from Brazil.

**3.**
**Javier:** Hi. My name's Javier. I'm from Argentina.
**Laura:** Nice to meet you, Javier. My name's Laura. I'm from Canada.

**4.**
**Sunan:** Hello. My name's Sunan.
**Andrew:** Nice to meet you, Sunan. I'm Andrew. Where are you from?
**Sunan:** I'm from Thailand. And you?
**Andrew:** I'm from Australia.

### GRAMMAR p. 4

1. They're in Tokyo.
2. I'm Canadian.
3. He's Chinese.
4. She's from Turkey.
5. We're American.
6. You're in my English class.

### REAL-WORLD LISTENING p. 7

**Mike:** Hi, I'm Mike. I'm from Miami. I work at a large computer software company. I love working here. I love my job, we work hard, but we have a lot of fun. The people are great! Come on, I'll introduce you to them!

**Mike:** This is Ben, he's a sales assistant. He's from Canada. Hi, Ben.
**Ben:** Hey, Mike. How are you?
**Mike:** Real good.

**Mike:** This is Olga.
**Olga:** Hi, Mike.
**Mike:** Hi, Olga.
**Olga:** Have you met Maria? She's new.
**Maria:** Nice to meet you.
**Mike:** Nice to meet you, too.

**Mike:** This is the marketing department. They're very busy.

**Mike:** Good morning, Kate.
**Kate:** Good morning, Mike.
**Mike:** Kate's our office manager. She's helpful.

**Mike:** Hi, Nancy, how are you?
**Nancy:** Fine.
**Mike:** Where are you from, Nancy?
**Nancy:** Well, I'm American but my family's from China.
**Mike:** Do you speak Chinese?
**Nancy:** A little bit.
**Mike:** Oh, that's nice. Well, have a nice day.
**Nancy:** Thanks.

**Mike:** Hi, Yasuko.
**Yasuko:** Hi, Mike.
**Mike:** You're from Japan, right?
**Yasuko:** I am. I'm from Tokyo.
**Mike:** So you speak Japanese?
**Yasuko:** I do. I'm bilingual. I speak both English and Japanese.
**Mike:** That's great. Okay. Well, I'll see you later.
**Yasuko:** OK, bye, Mike.

**Chris:** Hey there. I'm Chris from Boston. I'm the office assistant. I really like working here, but today's really busy.

**Kelly:** Hi, I'm Kelly. I'm the receptionist. I'm Canadian. I'm from Toronto.

**Josh:** Hi, I'm Josh. I'm from New York, and I'm a designer.
**Mike:** As you can see, people here are from many different countries.

**Mike:** So that's everybody. Aren't they nice? I'll see you around the office!

## Unit 2 >>>>>

### GRAMMAR p. 10

1. We get up at seven o'clock.
2. He cooks dinner.
3. I go to work at nine o'clock.
4. She finishes work at five.
5. They eat a big lunch.
6. I go to bed at eleven o'clock.

### REAL-WORLD LISTENING p. 13

**N:** This is Indojit. He lives with his family in the Thar Desert. His mother and father keep goats and cows. Early in the morning, Indojit gets up and milks the animals. Indojit's mother and sister also work early in the morning.

They walk to get water. It is a long trip—several miles. There isn't a lot of water in the desert, so it's very important. At ten o'clock, Indojit's father starts work and Indojit walks to school. He has English class at 10:30. It's his favorite. In the late afternoon, Indojit finishes school and comes home. He sometimes takes a camel! At home, his mother cooks dinner. She cooks with spices and other things from trees and plants in the area. So she doesn't usually go grocery shopping. Indojit comes home from school and the family eats dinner together. They are all hungry at the end of the day, so they have a big dinner. They don't have a TV, so they sit and talk about their day. After dinner, Indojit sometimes helps his father with his work. His father makes shoes. After that, he does his homework, plays with his brothers and sisters, and goes to bed.

# Unit 3 >>>>>

## VOCABULARY p. 15

**Maria:** My name's Maria. What do I do in my free time? Well, I swim and I play tennis. I also play the violin.

**Joshua:** My name's Joshua. Free time? Let me see. I go to the gym every day, and I play the piano. Oh, and I play soccer on weekends.

**Anna:** My name's Anna. In my free time I… hmm… I run every morning. In the evenings, I usually visit with friends. What else… OK, I watch TV… a lot of TV.

## REAL-WORLD LISTENING p. 19

**Sarita:** My name's Sarita Lou. I'm from California and I live in Brooklyn, New York. I'm a professional dancer and during the day I dance and in my free time I teach yoga. I love yoga because it's good for your health and it makes people happy. Yoga is a leisure time activity, but it's also a challenging sport. On an average day, I wake up at 8:00 am and I drink some juice. I go to the subway and go to the dance studio. I do modern dance, I don't do ballet. I do jazz dance. I also do West African dance and hip hop. I have a dance DVD that I sell to help people exercise. It's called *Dance and Be Fit*, *Hip Hop Cardio*. I practice, I sometimes perform so after work, I go uptown to the yoga school and I teach my yoga class. In my typical yoga class I have six or seven students. In the beginning of class, we sit and breathe, then we stand and stretch. We stretch for 30 minutes, then we sit down and we stretch on the ground. My favorite part of class is the end of class. We sit, relax, breathe and it feels really good. So after teaching yoga, I'm usually pretty tired, so I go home and make dinner. After dinner, sometimes I watch a movie or old TV shows before I go to bed. I don't have a lot of free time, but when I do, I try to socialize with friends. I feel lucky that I teach yoga and that I dance every single day. It makes me feel amazing and I'm really healthy. I get tired from teaching all the time and dancing all the time, but I don't work in an office and I love what I do, so I feel pretty lucky.

# Unit 4 >>>>>

## REAL-WORLD LISTENING p. 25

**Donna:** My name's Donna. I am married; I have a husband, Lenny. I have four kids. Today we are having a birthday party for my daughter, Alexa. She is turning 12. Grandma and Grandpa are coming, we're just going to have a good time today. I have three boys and I have one girl. My daughter, Alexa, she has a lot of friends. She's just a happy little girl. My oldest one's 19. His name's Taylor. Then I have one that's 15, that's Brandon. He likes to skateboard. I have a son that's 14, that's Jordan. He plays basketball. He plays baseball and he plays football. And my husband, Lenny, he works for a pharmaceutical company and he loves all sports. He coaches a lot of teams. He coached basketball for my son and his friends and his favorite is baseball. What do I like to do? Well, I work. I am a hairdresser. I like to garden. I like to cook. I like to read. I have one brother and I have a sister. My sister Lisa is coming today with my niece and my nephew, Alexis and Ryan, and my children just love to play with them. My parents I see about maybe once a month. That's Grandma and Grandpa and they don't live here. They live about an hour or so away. My father has three brothers and three sisters. And my mother has one sister and three brothers. I don't see my aunts and uncles that often. They all live pretty far away in different states. My mother-in-law, she lives very close by and we do see her very often. My husband has a brother and a sister, my brother-in-law and sister-in-law were not able to make the party today. My parents love spending time with their grandchildren and my children have lots of fun with their grandparents.

# Unit 5 >>>>>

## REAL-WORLD LISTENING p. 35

**N:** This town in Florida calls itself "The Villages." There are more than 20,000 people here, and it seems like a normal town. There's a pretty square, parks, nice stores, and more. The only difference? Everyone here is 55 or older. The average age is in the late 60s, and no one under 18 can live here. There are no schools, but the town has its own TV station. Sharon Shrepple is 63. She lives in the town and she's a reporter for the station. She's talking to people at a recreation center. Today, there's a dance class here. People also come to socialize and play cards. They even swim and play tennis here.

**Sharon:** I'm trying to get the hang of this, but I'm not too sure.

**N:** Life here is wonderful for elderly people. It's sunny 90% of the time. And after years of work, they finally have free time to enjoy it. Florida is home to millions of retired people. Every year, there are more people over 65. In fact, in 30 years, there will be more old people than young people in the US. There is a gym at The Villages. People here want to stay healthy and fit in their 60s, 70s, 80s, and beyond. Towns like The Villages are the future. Every year, there are more towns like this in Florida, in the U.S., and in many other countries around the world.

# Unit 6 >>>>>

## REAL-WORLD LISTENING p. 41

Mongolia. A land of extremes. More than 1,000,000 people live on this land. Many are nomads. They move from place to place. Some of them move 10 times a year.

It is a huge country, the size of Western Europe, and over one thousand five hundred meters high. When Mongolians move, they take their houses with them. They live in these white tents, called gers. Gers are perfect for long, cold Mongolian winters. Mongolians keep horses, buffalo, and sheep, even camels. When the animals move, the people take down their tents and move with them. They can put up the tents very quickly—in less than an hour. Everyone is welcome in a Mongolian ger. You don't even have to knock. One side of the ger is always open, so people can sit on carpets and visit with friends. They enjoy a healthy drink made from horse's milk. They also make cheese and butter. Sheep are very important in Mongolia. They provide meat and wool. Mongolians use the wool to make the roof and walls of the gers. The people waste nothing here. The fuel for this oven comes from the horses. In the summer, it is warm again. Outside the tents, there are festivals and sports. The Mongolian way of life continues.

# Unit 7 >>>>>

## VOCABULARY p. 43

1.
**A:** What date is our meeting?
**B:** It's on April sixteenth.
2.
**A:** When is our English test?
**B:** I don't know the date, but it's in May.
3.
**A:** When is your dentist's appointment, Kate?
**B:** It's on September twentieth.
4.
**A:** Hello. Is Helen there, please?
**B:** No. I'm sorry. She's away until October seventh.
5.
**A:** When is your wedding anniversary, Sam?
**B:** It's on December first.
6.
**A:** When's John's birthday?
**B:** It's in February, but I don't know the date.

## REAL-WORLD LISTENING p. 47

**Emiliano:** Hi, I'm Emiliano. I'm 23 years old from New York City, and this is my audition tape. I can play guitar, I can sing, and today I want to play some songs for you. I can play the blues. Or I can play some Cuban music. And some flamenco—I can play some flamenco.

No way, I can't dance at all. Man, music is my life. I play all day at home, outside. Sometimes I practice all night long. All my friends are musicians. So we're always playing, but when we're not, we're listening, and we go to concerts together. This is one of my songs and it's called "I Can't Have." "You are the only girl I dream of, the only girl I speak of; you're what I can't have. I know that my mind should move on, groove on, find a new one, but I just can't shake you. You're what I can't have. I can't have all I ever want. I can't have all I ever need. All I want, all I dream, I never get the loving from you that I need." Well that's my music. I hope you like it. I know I'm the best choice for the show. So pick me. I can't wait. "Loving you is all I want to do. It's all I want to do. It's all I want to do." Pick me.

# Unit 8 >>>>>

## GRAMMAR p. 50

He repairs computers.
We wash dishes.
They work in an office.
We study English.
She plays soccer.
I cook dinner.

## REAL-WORLD LISTENING p. 53

**N:** Johannes Scheffel is a helicopter pilot in the mountains of Austria. Tourists come here from all over the world. But Johannes doesn't take tourists here, or any passengers. He brings food and other things to hotels and restaurants in the mountains. His helicopter can carry more than 2000 pounds. And it can fly at 150 miles per hour. The pilots start work in late April, when there is less snow. Today, Johannes is bringing a food order to this ski hotel. Even now, the snow is still everywhere. The helicopter leaves. The manager and his chef are putting away the supplies.

**N2:** In the past, this took 3 or 4 hours. Now, everything is here right away. It's so easy.

**N:** The heli team is moving on to the next job. They do more than 15,000 deliveries a year. This is the main office, about 200 miles away. This man is scheduling helicopters for the company. The company schedules up to 40 helicopter trips in a day, and more than 300 in a week. The company is training these pilots for the job. The training isn't easy. It takes more than 1,000 hours of flying time. The pilots like their jobs, but there are some problems. It is dangerous work, with a lot of stress. It can also be bad for their health.

**N3:** In one day we go up and down 80,000 feet, maybe more. After a long day, you can really feel it.

**N:** Today, Johannes Scheffel and his coworkers brought more than 19 tons of food and drink to the mountains. Another day's good work for man and machine.

# Unit 9 >>>>>

## VOCABULARY p. 59

1. (Sound effects): A soccer game: fans cheering, the sound of a soccer ball being kicked
2. (Sound effects): A concert: classical music
3. To be or not to be. That is the question.
4. **Male:** Oh, I like this picture. Who's it by?
   **Female:** Picasso.
5. **Customer:** Do you have this shirt in a large?
   **Store clerk:** Let me see… yes, here's a large.
6. Coming soon to a theater near you… The Strangest Summer… A romance starring …

## REAL-WORLD LISTENING p. 63

Memphis, Tennessee. Rock and roll was born here, at Sun Studio. Old and young, modern and traditional. It's all here by the Mississippi River. For great concerts and nightlife, head to the clubs on Beale Street. There's the blues…rock and roll…With 4 million visitors a year, it's always crowded but never boring. The Beale Street restaurants are all about barbecue. It's delicious! There are more than 25 clubs and restaurants on Beale Street, and they're open until 5:00 a.m. The Peabody Hotel is in an ideal location. You can walk to Beale Street, visit with the locals, or take in the view from a trolley car. The home of rock and roll and the blues, Memphis is an enormous part of music history and American history. The singer Elvis Presley lived here for many years. His house, Graceland, is now a museum. The Memphis Belle was an airplane used in World War Two. The Lorraine Motel. Martin Luther King, the great African American leader, was tragically killed here in 1968. The hotel is now the National Civil Rights Museum. Home of the blues… by the mighty Mississippi… Memphis, Tennessee. An American city.

# Unit 10 >>>>>

## VOCABULARY p. 65

**Linda:** So, is this a photo of your class at school, Adam?
**Adam:** Yes, it is. That's me there with my friend, Martin. That's Mrs. Harris. She's our science teacher. She's really nice. I like science. It's my favorite subject.
**Linda:** What other subjects do you like?
**Adam:** Oh, well, I like math. But I don't like history. It's so boring!
**Linda:** And what about you, Holly? What's your favorite subject?
**Holly:** Well, I like gym, but my favorite subject is Spanish.
**Linda:** Oh, are you good at languages?
**Holly:** Yes, I am.
**Linda:** I was good at languages, too, when I was in school—Spanish and Japanese, but my favorite was Portuguese.
**Holly:** Portuguese! Wow! Cool! We don't have that at our school.
**Linda:** And which subjects don't you like?
**Holly:** I don't like science.
**Linda:** So, what do you like, Joe?
**Joe:** I like history and I like art, too—oh, and P.E.
**Linda:** Well, you're good at sports.
**Joe:** Yeah, but history's my favorite subject. I don't like math. I'm not very good at it.

## REAL-WORLD LISTENING p. 69

**Alison:** Hey, I'm Alison King and this is Hawthorne High School. I'm a senior here, so I graduate in a few months. Finally! Yesterday was a typical day at Hawthorne. School started at 8:00—way too early. So I got up around 6:30. I got to school and I hung out in the hallway and chatted with friends. At 8:00 a.m., the principal made the morning announcements. The announcements are so boring. Nobody listens to them. The principal is like the boss of the school. We call him "sir." He loves it. Anyway, I had Spanish for first period. Hey look, that's me! We have eight periods at Hawthorne. They're 45 minutes each. For second period, I had science. That was fun…then math…boring…English…boring…Then history, which is awesome cause my friends are in it. I also have to take physical education. Ugh. Everyone calls it "gym," and everyone hates it. My classes in the afternoon are great, though. I'm really into them. There's music…and computer science. I love the Internet. Art class is great, too. We get to listen to music sometimes. Last year I had home economics. It was cool. You learn to cook and stuff. Fifth period is definitely my favorite class: lunch! A few years ago, the school food was really gross. But it's pretty delicious now. Lunch is awesome. You can hang out with your friends and just chat and relax. At 2:46, eighth period finishes and we can go home! Some people stay after school for sports, or for the school play and stuff. But not me! I love Hawthorne, but I am really ready to graduate!

# Unit 11 >>>>>

## REAL-WORLD LISTENING p. 75

We all love laughter. It feels good. But there's more—laughing is also healthy. Some doctors say to laugh 30 minutes a day. Babies first start to smile at the age of 4 weeks. It's the first time they can communicate with another person. Most of us started to laugh at around four months old. This is true for all children, in countries around the world. Even blind children. They can't see, but they start to smile and laugh at the same age. Smiles and laughter are an international language we can all understand. Scientists say laughter is good for your body and your mind. In this experiment, people watched a funny movie. They tested the people after the movie. Their health improved. Scientists say laughter is good for stress. It's also good for colds and the flu. So laughing really is good for us. The average person laughed a lot as a child—400 times a day. However, many adults laugh only a few times a day. Some never laugh. This can be very unhealthy. This hospital is using laughter to help patients. These "clown doctors" are not real doctors.

But the laughter is very real, and so are the health benefits. Illnesses are less painful. Patients get better faster. Laughter really is the best medicine.

# Unit 12 >>>>>

## VOCABULARY p. 77

A: Hello, Marco. Did you and Isabel have a good vacation?
B: Yes, we did, thanks. It was great.
A: Where did you go?
B: We went on a cruise in the Caribbean.
A: Wow, that sounds wonderful! Where did you get on the boat?
B: In Florida. We flew to Miami.
A: Was the flight OK?

B: Well, yes, but we had a problem. I forgot my driver's license!
A: Oh, no. When did you realize that?
B: When we were at the airport.
A: Really? So what did you do?
B: We called our neighbor. She went to the house, and drove to the airport with my license.
A: That wasn't a good start to a vacation.
B: No, it wasn't.
A: Did you miss the plane?
B: Yes, we did, but they put us on another plane, so it wasn't too bad, and we didn't miss the boat in Florida.
A: That's good!

## REAL-WORLD LISTENING p. 81

N: Honolulu, Hawaii is a city on the island of Oahu. In the Hawaiian language, "Oahu" means "gathering place." It's a good name. The people of Honolulu are from the U.S., Japan, Europe, and many other places. In the past, Hawaii had kings, like this one, Kamehameha. Then, in 1898, Hawaii became a part of the U.S. Waikiki is a famous beach in Honolulu with many popular hotels and shopping centers. There are more than 65,000 visitors a day. Visitors can also see traditional Hawaiian music. Once a week, the Royal Hawaiian Band gives a concert at the old royal palace. People here also do the hula, a traditional Hawaiian dance. The movements of the hula dance have special meanings. They can tell a story, or describe a famous king in history. Hawaii is also famous for big waves—and with the big waves, surfing. Surfing has a very long history here. Many surfers say Hawaii is the best place in the world for surfing. Honolulu is full of life, day and night. On the Fourth of July, Hawaiians celebrate American Independence Day. There are fireworks at night, and barbecues on the beach.

# Unit 13 >>>>>

## REAL-WORLD LISTENING p. 91

Joe: Hi, my name is Joe. I'm a food vendor and a waffle maker in New York City. We have a food truck. It's big and yellow, and we sell waffles and coffee and ice cream on the side of the street. Many of our customers are local business people that will come down on their lunch break and grab a waffle. Our waffle truck is parked right outside of their building. So, because they don't have much time they can just grab a waffle and get right back inside. Every day I go to our garage in Brooklyn and pick up the truck, load it up and drive it into the city where I park it on the corner and open up and sell waffles. Every day, we have to make sure we have about 16 boxes of strawberries; probably five big bunches of bananas, we have to have a big tub of maple syrup, a couple of things full of chocolate fudge. In the waffle mix, there's flour, eggs, milk, some secret spices, some vanilla. We have two kinds of waffles. We have the Brussels waffle, which is light and crispy, and we have the liege, which is soft, chewy, and a little bit sweeter. So then, when somebody orders a waffle, we write their order on the back. If it's strawberries, a spoonful of strawberries—we put about half of banana, whipped cream—you do two little circles of whipped cream. We have a special waffle right now that has peaches and whipped cream. We put some peaches on there, put some whipped cream, dash a little cinnamon sugar, and it's ready. Most people think waffles are only for breakfast. And they aren't traditionally in Belgium eaten for breakfast. We have some really great lunch waffles. We have an Italian cheese waffle. We have a fried chicken waffle. On the streets in New York now you can get any sort of food that you could want. There are Middle Eastern trucks that serve falafel and chicken and rice. There are Mexican trucks that serve tacos and tortas and burritos. Chinese trucks that serve dumplings, Korean trucks that have lunch boxes like they do in Korea. And now thanks to us, traditional Belgian waffles are becoming very popular in the United States. My favorite part of working in this food truck is it makes people happy. We don't have angry customers. When people come to the truck, they're happy to be there, and that makes me happy.

# Unit 14 >>>>>

## VOCABULARY p.93

Do you like this shirt?
Do you like these pants?
Do you like these shoes?
Do you like this hat?
Do you like this dress?
Do you like these jeans?

## REAL-WORLD LISTENING p. 97

The fashion business is always changing and always colorful. Today's traditional clothing becomes tomorrow's trendy new thing. Fashion has a long history, and it started with plants and animals. These people in Polynesia still make fabric from a plant called "flax." It's strong, it's waterproof, and it's free. Perhaps the most famous plant in fashion is this. Cotton. Cotton's in everything from T-shirts to socks. And it is used to make denim for jeans. Fashion also owes a lot to silkworms. These insects are big business.

One silkworm can make a thread three miles long. After a lot of very difficult work, the result is gorgeous, luxurious silk. Then there are sheep. One sheep has enough wool to make a big, warm sweater. In the Himalaya mountains, people keep warm with wool from the yak. Making fabric isn't easy. It takes skill and training and sometimes special equipment. In many cultures, fabric can have special meanings. The patterns can show where you come from, or your family's history, like these fabrics in Scotland. In India, people still make patterns on fabric in the traditional ways. These traditional ways take more time but look beautiful. There are also modern factories. This kind of pattern, paisley, takes hours and hours of work. Every pattern is different. Today, computers are a big help. From traditional to modern…from dark denim to shiny gold… a world of fashion and fabrics.

# Unit 15 >>>>>

## REAL-WORLD LISTENING p. 103

There are deserts in the north and mountains in the south, but the heart of Chile is the capital, Santiago. One-third of the country lives here. There is an interesting mix of modern skyscrapers and older buildings. The Plaza de Armas. The center of the plaza is Kilometer Zero. Every distance in Chile is measured from this point. About 1,000 kilometers north of Santiago is the Atacama Desert. No place is drier. Few people live here. In some parts of Atacama, there is only 1 millimeter of rain every year. It's like nothing on Earth. In fact, some scientists say it is more similar to Mars. In the Atacama Desert, a guide can take you to the Valley of the Moon. Sunset is a good time to go. The weather is cooler, and the views are more beautiful then. To the south of Santiago is a totally different world: Patagonia. At over a million square kilometers, Patagonia is huge, ten times bigger than the Atacama Desert. Here, you can experience a very different Chile. It's greener here. The weather is cooler and rainier and also more unpredictable. If you like riding horses, here is the place. In this peaceful place, the horses are even more relaxed than you. The Gray Lake Glacier is a must-see. As it slowly melts into the lake, the ice has a beautiful blue color. Chile has wonderful natural beauty and a fine capital city at its heart. The memories you make here will last a lifetime.

# Unit 16 >>>>>

## YOUR STORY p. 108

**Peter:** Excuse me, where does the flight to Rio leave from?
**Clerk:** Gate 15. Security is right over there.
**Peter:** Thank you.
**Peter:** Sarah! Sarah!
**Sarah:** Peter? What are you doing here?
**Peter:** I'm trying to find you.
**Sarah:** I'm going back to Brazil.
**Peter:** I know. I feel terrible about everything.
**Sarah:** It's going to be okay. She just needs to be in the hospital.
**Peter:** Hospital? Who?
**Sarah:** My sister. She broke her arm. That's why I'm going home.
**Peter:** Oh, I see. I thought…Well, my ex-girlfriend was in town, and I thought…
**Sarah:** Oh? Well, as long as she is an ex-girlfriend, it doesn't bother me.
**Peter:** What about you? Jordan told me you had a boyfriend back in Brazil.
**Sarah:** I think he's going to be an ex-boyfriend very soon.

## REAL-WORLD LISTENING p. 109

**Josh:** Hi, I'm Josh. I'm a tour guide here in New York City. "Welcome everyone to beautiful New York City!" On my tour, I show people the most famous places in New York, places like the Empire State Building, Times Square, Battery Park, Statue of Liberty, SOHO, Greenwich Village, Chinatown, Little Italy—everything in between. "This is where we invented the hamburger. This is where we invented the hotdog. And that's pretty much it." I love being a tour guide. It's the best job in the world because I get to meet so many great people from all over the place. I get to meet people from Afghanistan to Algeria to Italy to France, Germany… "Let's do some interviews from some of our international travelers. Where are you from, sir? Birmingham, England. Fantastic! So, welcome to New York City, sir. What's your favorite part about New York so far? Empire State Building, that's a good answer; I'm afraid that answer is wrong. We were looking for my tour guide on the downtown tour." The Empire State Building is the tallest building in New York City. The Empire State Building is 102 stories tall. I think the only thing more popular than the Empire State Building: Statue of Liberty. Times Square is the most crowded part of New York City. People come from all over the world to visit us here in New York, and where do they stay? In Times Square. "Up on the left of the bus, we have Macy's, the world's largest store." Macy's is one of the oldest stores in America. They've been around since 1858. Greenwich Village isn't just one of the oldest neighborhoods in New York City, it's also one of the coolest places in town, with the best nightlife. Manhattan has two rivers on either side of it. On the west side, you have the Hudson River. On the east side, the East River. We've got the Brooklyn Bridge. That's the oldest bridge in New York City. When it first opened, it was the longest bridge in the world and the tallest structure on earth. It's a mile and a half long, and you can actually walk across the Brooklyn Bridge, even today. I think the Flatiron Building is the most beautiful building in this city. New York City hot dog stands have the cheapest hot dogs in New York. Unfortunately, they're also the worst. You don't know how long those hot dogs have been sitting in that hot dog water. New York is the greatest city in the world, and I love working here. Listen to all that street noise—that's the real New York.

# Grammar Reference

## Unit 1 >>>>>

### be

| Positive statements | | | Negative statements | | |
|---|---|---|---|---|---|
| I | 'm | | I | 'm not | |
| He She It | 's is | here. | He She It | isn't (is not) | here. |
| We You They | 're are | | We You They | aren't (are not) | |

**NOTE:** We don't say ~~I ain't.~~

| Questions and short answers | | | | | | | | | |
|---|---|---|---|---|---|---|---|---|---|
| Am | I | | | I | am. | | I | 'm not | |
| Is | he she it | here? | Yes, | he she it | is. | No, | he she it | isn't | here. |
| Are | we you they | | | we you they | are. | | we you they | aren't | |

**NOTE:** We don't use short forms in positive short answers. NOT Are they here? ~~Yes, they're.~~

## Unit 2 >>>>>

### Present simple—statements

We use the present simple for:

1. permanent states: I **work** in a shop. They **speak** Spanish.
2. regular activities: I **wake up** at 6 a.m. They **don't walk** home.

| Positive statements | | | Negative statements | | |
|---|---|---|---|---|---|
| I We You They | work | here. | I We You They | don't (do not) | live work study |
| He She It | works | | He She It | doesn't (does not) | |

**NOTE:** After *doesn't* we use the base form, not the third person form of the verb.

She doesn't live here. NOT ~~She doesn't lives.~~

### Adverbs of frequency

never — sometimes — often — usually — always
0% ← → 100%

1. Adverbs of frequency go before a verb (but not the verb *be*):

    I **never** wake up early on Sunday.   He **always** starts work at nine o'clock.

2. Adverbs of frequency go after the verb *be*:

    The traffic is **always** very bad.   She is **sometimes** late in the morning.

120

# Unit 3 >>>>>

## Present simple—questions

| yes/no questions and short answers | | | | | | | | |
|---|---|---|---|---|---|---|---|---|
| Do | I you we they | play tennis? | Yes, | I you we they | do. | No, | I you we they | don't. |
| Does | he she it | | | he she it | does. | | he she it | doesn't. |

**NOTE:** After *does* we use the base form, not the third person form of the verb.
> *Does he play tennis?*
> NOT *Does he plays tennis?*

**NOTE:** We don't use the verb in short answers.
> *Yes, I do./Yes, he does.*
> NOT *Yes, I play./Yes, he plays.*

### *Wh-* questions

We use the same word order for *yes/no* questions and *Wh-* questions.

> *Do you read books?*
>> *What books do you read?*
>
> *Do you play tennis?*
>> *Where do you play?*

# Unit 4 >>>>>

## have/has

| Positive statements | | |
|---|---|---|
| I You We They | have don't have (do not have) | children. |
| He She It | has doesn't have (does not have) | |

| Questions and short answers | | | | | | |
|---|---|---|---|---|---|---|
| Do | I you we they | have | a sister? | Yes, | I you we they | do. |
| | | | | No, | I you we they | don't. |
| Does | he she (it) | | | Yes, | he she it | does. |
| | | | | No, | he she it | doesn't. |

**NOTE:** We ask questions about numbers with *how many*.
> ***How many** sisters do you have?*

GRAMMAR REFERENCE 121

## Unit 5 >>>>>

### there is/there are

1. We use *there is/there are* to describe a scene or a place.
2. We usually use *a/an* or *some/any* after *there is/are*.

| Positive statements | Negative statements |
|---|---|
| There's **a** cash machine at the station. There are **some** shops over there. | There isn't **an** internet cafe in town. There aren't **any** parks near here. |

**NOTE:** We don't use short forms in positive short answers.

*Is there a hotel on this street?*

*Yes, there is.* NOT ~~Yes, there's.~~

*Are there cafes near here?*

*Yes, there are.* NOT ~~Yes, there're.~~

## Unit 6 >>>>>

### Present continuous

We use present continuous to say what is happening now.

| Statements | | |
|---|---|---|
| I | 'm (am) | drinking coffee. reading. watching TV. |
| He She (It) | 's (is) isn't (is not) | |
| We You They | 're (are) aren't (are not) | |

| Questions and short answers | | | | | | |
|---|---|---|---|---|---|---|
| Am | I | | Yes, | I | | am. |
| | | | No, | | | 'm not. |
| Is | he she (it) | drinking coffee? reading? watching TV? | Yes, | he she (it) | | is. |
| | | | No, | | | isn't. |
| Are | we you they | | Yes, | we (you) they | | are. |
| | | | No, | | | aren't. |

**NOTE:** We don't use short forms in positive short answers.

*Are they watching TV?*

*Yes, they are.* NOT ~~Yes, they're.~~

| Spelling: 3rd person singular | | |
|---|---|---|
| | Base | Participle |
| Most verbs | talk | talking |
| Verbs + -e | make | making |
| Verbs + short vowel + one consonant | put sit | putting sitting |

# Unit 7 >>>>>

## can/can't

| Statements | | |
|---|---|---|
| He She (It) We You They | can can't (cannot) | swim. speak Chinese. use a computer. |
| NOT ~~I can to swim.~~ | | |

| Questions and short answers | | | | | | |
|---|---|---|---|---|---|---|
| Can | I we you they he she (it) | play today? | Yes, No, | I we you they he she (it) | can. can't. | |

# Unit 8 >>>>>

## Present simple and present continuous

### Present simple

1. We use the present simple tense for regular activities:

    I usually **wake up** early.

    He **doesn't walk** to work.

    She **often works** in the morning.

2. We often use the present simple tense with these time expressions:

    **always**

    **usually**

    **often**

    **never**

    **every day**

    **on weekends**

    **on (Wednesdays)**

### Present continuous

1. We use the present continuous tense to talk about what is happening now:

    I'**m having** lunch in a restaurant today.

    He'**s listening** to his CDs at the moment.

    They **aren't working** in London this week.

2. We often use the present continuous with these time expressions:

    **now**

    **at the moment**

    **today**

    **tonight**

    **this morning**

    **this week**

GRAMMAR REFERENCE 123

## Unit 9

### Past simple—to be

| Statements | | |
|---|---|---|
| I<br>He<br>She<br>(It) | was<br>wasn't<br>(was not) | at the shopping mall yesterday. |
| We<br>You<br>They | were<br>weren't<br>(were not) | |

### Questions: word order

**He was** at a football game.

**Was he** at a football game?

| Questions and short answers | | | | | | |
|---|---|---|---|---|---|---|
| Was | I<br>he<br>she<br>(it) | at home yesterday? | Yes,<br>No, | I<br>he<br>she<br>(it) | was.<br>wasn't. |
| Were | we<br>you<br>they | | Yes,<br>No, | we<br>you<br>they | were.<br>weren't. |

We often use the past simple with these time expressions:

**yesterday, last week, last night, last month, last year, yesterday evening**

## Unit 10

### Past simple statements—positive

| Regular verbs | Present | Past |
|---|---|---|
| Most verbs, add -ed | work | worked |
| Verbs + -e, add -d | like | liked |
| Verbs + short vowel + one consonant | stop | stopped |
| Consonant + -y | study | studied |

**NOTE:** When a verb ends with vowel + -y we just add -ed.

*play    played    enjoy    enjoyed*    BUT    *study    studied*

The past simple is the same for all subjects.

*I liked    He liked    We liked    They liked*

### Past simple statements—negative

| Statements | | |
|---|---|---|
| I, He, She, (It),<br>We, You, They | didn't<br>(did not) | go out.<br>like the movie. |

The past simple negative is the same for regular and irregular verbs and all subjects.

**NOTE:** We use the base form, not the past simple form, after *didn't*.

*I didn't **work**... NOT I didn't worked...*

# Unit 11 >>>>>

## Past simple—questions

1. We use the same form for regular and irregular verbs in past simple questions.

| Questions and short answers | | | | | | |
|---|---|---|---|---|---|---|
| Did | I<br>we<br>you<br>they<br>he<br>she<br>(it) | study?<br>go skiing? | Yes,<br><br>No, | I<br>we<br>you<br>they<br>he<br>she<br>(it) | did.<br><br>didn't. | |

2. We use the same form for all subjects.

   ***Did** she **go** out? No, she **didn't**.*
   ***Did** they **go** out? No, they **didn't**.*

3. We don't use the past tense form in questions.

   *Did you go swimming?* NOT ~~Did you went swimming?~~

4. We use the question form in *Wh-* questions, too.

   |  | ***Did you do*** *that on the weekend?* |
   |---|---|
   | When | ***did you do*** *that?* |
   |  | ***Did you go*** *skiing?* |
   | Where | ***did you go?*** |

# Unit 12 >>>>>

## be + going to

We use *be + going to* to talk about intentions or plans for the future.

| Statements | | | |
|---|---|---|---|
| I | 'm (am)<br>'m not (am not) | going to | play tennis.<br>have a party. |
| He<br>She<br>(It) | 's (is)<br>isn't (is not) | | |
| We<br>You<br>They | 're (are)<br>aren't (are not) | | |

## be + going to—questions and short answers

We make questions and short answers with *be + going to* like this:

***She's*** *going to get up early.*

***Is she*** *going to get up early? Yes, she is. / No, she isn't.*

***You're*** *going to get up early.*

***Are you*** *going to get up early? Yes, I am. / No, I'm not.*

NOTE: We don't use short forms in short answers.

*Are you going out? Yes, I am.* NOT ~~Yes, I'm.~~

# Unit 13 >>>>>

## Count and noncount nouns

1. Some nouns are count. They have a singular and a plural form. We use *a/an* with the singular and *some/any* with the plural.

| Singular | Plural |
|---|---|
| I have **a** pen. | I have **some** pens. |
| I don't have **a** pen. | I don't have **any** pens. |
| Do you have **a** pen? | Do you have **any** pens? |

2. Some nouns are noncount. They don't have a singular and plural form. We don't use *a/an* with these words.

| Noncount nouns | |
|---|---|
| I have **some** bread.<br>NOT I have a bread.<br>OR I have some breads. | I don't have **any** bread.<br>Do you have **any** bread? |

3. These words are usually noncount.

> Drinks and other liquids: *coffee, water, wine, gas*
>
> Food which you only eat part of: *meat, fish, bread, cheese, chocolate, fish*
>
> Things which you only use part of: *toothpaste, shampoo, soap, makeup*
>
> Some general words: *time, music, medicine, paper, money, information*

## How much/how many?

| Count nouns | Noncount nouns |
|---|---|
| **How many** cookies do you eat? | **How much** cheese do you eat? |
| I don't eat **many** cookies. | I don't eat **much** cheese. |
| You eat **a lot of** cookies. | You eat **a lot of** cheese. |

# Unit 14 >>>>>

## Adjectives

1. Adjectives describe a noun. They go before the noun in phrases.

> a **blue** shirt     **new** shoes NOT a shirt blue     shoes new

Adjectives go after the verb *be* in sentences.

> *The shirt is **blue**.*
>
> *The shoes are **new**.*

2. Adjectives are the same for singular and plural nouns.

> *an old car*
>
> *old cars*     NOT olds cars

3. We can modify an adjective with *very*.

> *That's a **very** expensive car.*
>
> *Those jeans are **very** nice.*

4. When there is more than one adjective, they follow this order:

| Adjective order | | | |
|---|---|---|---|
| 1. quality | 2. size | 3. age | 4. color |
| good<br>bad<br>beautiful<br>nice | big<br>small<br>long<br>short | old<br>new | black<br>white<br>blue<br>red |

> *a new black dress, a nice long vacation, a small white car*

# Unit 15 >>>>>

## Comparatives

1. We use the comparative form of an adjective to compare two things.
2. To form most comparatives, we add *-er* to the adjectives.

    cold    cold**er**    long    long**er**

3. Note these spelling rules:

|  | Base | Comparative |
|---|---|---|
| -e | nice | nicer |
| Short vowel + one consonant | wet | wetter |
| -y | windy | windier |
| Irregular | good<br>bad<br>far | better<br>worse<br>further |

4. For most adjectives with two or more syllables (except where the second syllable is –y), we use **more**.

    interesting    **more** interesting
    famous         **more** famous

    BUT    happy    happ**ier**

5. To compare two things, we use a **comparative + than**.

    Tomorrow is going to be **better than** today.
    Canada is **bigger than** the United States.

# Unit 16 >>>>>

## Superlatives

1. To form most superlatives, we add *-est* to the adjectives.

    cold    cold**est**    long    long**est**

2. We use *the* to make a superlative.

    I'm **the** oldest person here. NOT ~~I'm oldest person here.~~

3. The spelling rules are the same as for the comparative.

|  | Base | Comparative | Superlative |
|---|---|---|---|
| -e | nice | nicer | the nicest |
| Short vowel + one consonant | wet | wetter | the wettest |
| -y | windy | windier | the windiest |
| Irregular | good<br>bad<br>far | better<br>worse<br>further | the best<br>the worst<br>the furthest |

4. For most adjectives with two or more syllables (except where the second syllable is -y), we use **the most**.

    interesting    more interesting    **the most** interesting
    famous         more famous         **the most** famous

5. After a superlative, we usually use *in*.

    Everest is the highest mountain **in** the world.

# Word List

## Unit 1 >>>>>

Argentina
Australia
Brazil
Canada
China
Japan
Korea
Mexico
Portugal
Spain
Thailand
Turkey
the United States

American
Argentinian
Australian
Brazilian
Canadian
Chinese
Japanese
Korean
Mexican
Portuguese
Spanish
Thai
Turkish

### Challenge words
assistant
bilingual
busy
delighted
manager
multilingual

### Expressions
Bye.
Good afternoon.
Good evening.
Good morning.
Goodbye.
Goodnight.
Hello./Hi.
See you later.

### Real-world listening
busy
designer
helpful
marketing
receptionist
reports

## Unit 2 >>>>>

eat lunch
finish work
get up
go to bed
go to work
watch TV

### Challenge words
earn
like
live
stay
teach
wait
walk
wash

### Expressions
Oh, right.
That's funny.
Yes, I know.

### Real-world listening
camel
desert
goats
grocery shopping
plants
spices

## Unit 3 >>>>>

do yoga
go to the gym
go to the movies
play soccer
play the piano
visit with friends

### Challenge words
average
exercise
leisure
play computer games

128

socialize
typical

### Expressions
Can I take a message?
I'm sorry, she's out.
Who's calling, please?
Yes, please.
Please ask her to call me back.
Oh, well, can I leave a message, please?

### Real-world listening
ballet
breathe
cardio
fit
jazz dance
modern dance
socialize
stretch

# Unit 4 >>>>>
aunt
brother
brother-in-law
children
cousin
daughter
father
grandfather
grandmother
grandparents
great-grandparents
husband
mother
nephew
niece
parents
sister
son
uncle

### Challenge words
celebration
hometown
keep in touch
relatives

### Expressions
How about soup?
Let's get a sandwich here.
Why don't we get something to eat?

Good idea.
Thanks.
Yes, OK.

### Real-world listening
garden
hairdresser
have a good time
pharmaceutical company
skateboard

# Unit 5 >>>>>
ATM
bank
bus stop
cafe
drugstore
gas station
grocery store
gym
hair salon
hotel
park
parking lot
public restrooms
restaurant
square
subway station

### Challenge words
across from
at the end (of)
next to
on the corner (of)

### Expressions
Go down.
Go past.
It's on the left.
It's on the right.
Take the first left.
Take the first right.
Turn left.
Turn right.

### Real-world listening
get the hang of
healthy
play cards
recreation center
retired

**WORD LIST    129**

## Unit 6 >>>>>

armchair
bathroom
bathtub
bedroom
cabinet
carpet
curtains
dishwasher
dresser
kitchen
living room
microwave
mirror
refrigerator
sink
sofa
stove
table

### Challenge words
excited
putting together
putting up
surprise
taking a break
taking down

### Expressions
Let me give you a hand.
Thanks a lot.
That would be great.
Would you like some help?

### Real-world listening
buffalo
extremes
fuel
huge
knock
nomads
tents
wool

## Unit 7 >>>>>

January
February
March
April
May
June
July
August
September
October
November
December
first
second
third
fourth
fifth
sixth
seventh
eighth
ninth
tenth
eleventh
twelfth

### Challenge words
barbecue
celebration
make it
out of town
special occasion
volleyball

### Expressions
I can't explain.
I don't understand.
I'm only trying to help.
I'm worried about my interview.
What's the problem?

### Real-world listening
audition
flamenco
groove
guitar
musicians
the blues

## Unit 8 >>>>>

chef
doctor
flight attendant
mechanic
server
store clerk

### Challenge words
away from home

bus driver
client
look around
real estate agent
retire
tourist

### Expressions
Can/Could I borrow your hammer, please?
Can/Could you hold the nail, please?
Help yourself.
Of course.

### Real-world listening
deliveries
helicopter
passengers
pilot
stress
tourists

## Unit 9 >>>>>
art gallery
concert
movies
shopping mall
soccer game
theater / a play

### Challenge words
clubs
crowded
delicious
enormous
location
modern
nightlife
view

### Expressions
I'm so sorry to hear that.
That's terrible.
What's the matter?

### Real-world listening
crowded
enormous
modern
rock and roll
traditional
trolley car

## Unit 10 >>>>>
art
foreign languages
history
math
physical education (P.E.)
science

### Challenge words
accident
injure
nervous
remember
straight-A
tutor
video game designer

### Expressions
Congratulations!
How wonderful!
That's good.
That's terrific.

### Real-world listening
announcements
awesome
chat
graduate
gross
hang out
home economics
principal

## Unit 11 >>>>>
ankle
arm
back
chest
ear
elbow
eye
face
finger
foot/feet
hair
hand
head
knee
leg
mouth

neck
nose
shoulder
stomach
toe
tooth/teeth

### Challenge words
ambulance
ouch
painful
tripped
twisted

### Expressions
Do you have anything for (a cold/a headache/etc.)?
Is it for you?
Take two capsules every 12 hours.

### Real-world listening
blind
communicate
experiment
illness
improved
laughter
patients
smile

## Unit 12 >>>>>
go backpacking
go camping
go on a business trip
go on a cruise
luggage
passport

### Challenge words
a while
go backpacking
graduate
improve
ski resort
take it easy

### Expressions
I don't believe it!
Oh my goodness.
Well, what do you know!
You're kidding!

### Real-world listening
fireworks
hula
island
kings
movements
surfing
visitors
waves

## Unit 13 >>>>>
apples
beef
bread
carrots
cheese
chicken
eggs
fish
lettuce
milk
potatoes
tomatoes

### Challenge words
cereal
fried
grapes
healthy
herbs
hot chocolate
hungry
jam
muffin
noodles
on the way
toast
yogurt

### Expressions
Are you ready to order?
Can I have a blueberry muffin, please?
Could I also get some fruit and yogurt?
(Would you like) anything to drink?

### Real-world listening
cinnamon
customers
food truck
food vendor

garage
grab
waffle
whipped cream

## Unit 14 >>>>>

belt
blouse
boots
coat
dress
gloves
hat
jacket
jeans
pants
shirt
shoes
shorts
skirt
socks
tights

### Challenge words
badge
casual
cool
fashion model
photographer
scarf
sneakers
tie
T-shirt
uniform

### Expressions
How exciting!
Oh, really?
That sounds interesting.
That's great.
You look very nice.

### Real-world listening
cotton
denim
equipment
fabric
gorgeous
luxurious
pattern
trendy

## Unit 15 >>>>>

cloudy
cold
cool
foggy
hot
rainy
snowy
sunny
warm
windy

### Challenge words
American football
electronics
further
health insurance
salary
yard

### Expressions
Are you tired?
How did it go?
Welcome home.
When did you get back?

### Real-world listening
a guide
melts
memories
peaceful
scientist
skyscrapers
sunset
unpredictable

## Unit 16 >>>>>

bridge
cliff
field
forest
highway
hill
island
lake
mountain
ocean
railroad tracks
river
road

**WORD LIST** 133

tunnel
valley

### Challenge words
average
bungee jump
extreme sports
hike
jet boat
lovely
mountain bike
peaceful
scenery
waterfall

### Expressions
What are you talking about?
What do you mean?
What's the problem?
Why?

### Real-world listening
bridge
fantastic
interviews
neighborhood
tourist attraction
welcome aboard

# Irregular Verbs

| Base form | Past simple | Past participle |
|---|---|---|
| be | was/were | been |
| become | became | become |
| begin | began | begun |
| blow | blew | blown |
| break | broke | broken |
| bring | brought | brought |
| build | built | built |
| buy | bought | bought |
| can | could/was able to | been able to |
| catch | caught | caught |
| choose | chose | chosen |
| come | came | come |
| cost | cost | cost |
| cut | cut | cut |
| do | did | done |
| draw | drew | drawn |
| drink | drank | drunk |
| drive | drove | driven |
| eat | ate | eaten |
| fall | fell | fallen |
| feel | felt | felt |
| fight | fought | fought |
| find | found | found |
| fly | flew | flown |
| forget | forgot | forgotten |
| get | got | gotten |
| give | gave | given |
| go | went | been/gone |
| grow | grew | grown |
| have | had | had |
| hear | heard | heard |
| hit | hit | hit |
| hold | held | held |
| hurt | hurt | hurt |
| keep | kept | kept |
| know | knew | known |
| leave | left | left |

| Base form | Past simple | Past participle |
|---|---|---|
| let | let | let |
| lose | lost | lost |
| make | made | made |
| mean | meant | meant |
| meet | met | met |
| pay | paid | paid |
| put | put | put |
| read /rːd/ | read /rɛd/ | read /rɛd/ |
| ride | rode | ridden |
| ring | rang | rung |
| run | ran | run |
| say | said | said |
| see | saw | seen |
| sell | sold | sold |
| send | sent | sent |
| set | set | set |
| show | showed | shown |
| shut | shut | shut |
| sing | sang | sung |
| sit | sat | sat |
| sleep | slept | slept |
| speak | spoke | spoken |
| spend | spent | spent |
| stand | stood | stood |
| steal | stole | stolen |
| swim | swam | swum |
| take | took | taken |
| teach | taught | taught |
| tell | told | told |
| think | thought | thought |
| throw | threw | thrown |
| try | tried | tried |
| understand | understood | understood |
| wake | woke | woken |
| wear | wore | worn |
| win | won | won |
| write | wrote | written |

198 Madison Avenue
New York, NY 10016 USA

Great Clarendon Street, Oxford, OX2 6DP, United Kingdom

Oxford University Press is a department of the University of Oxford.
It furthers the University's objective of excellence in research, scholarship,
and education by publishing worldwide. Oxford is a registered trade
mark of Oxford University Press in the UK and in certain other countries

© Oxford University Press 2012

The moral rights of the author have been asserted

First published in 2012

2016 2015

10 9 8 7 6 5 4

**No unauthorized photocopying**

All rights reserved. No part of this publication may be reproduced, stored in a retrieval system, or transmitted, in any form or by any means, without the prior permission in writing of Oxford University Press, or as expressly permitted by law, by licence or under terms agreed with the appropriate reprographics rights organization. Enquiries concerning reproduction outside the scope of the above should be sent to the ELT Rights Department, Oxford University Press, at the address above

You must not circulate this work in any other form and you must impose this same condition on any acquirer

Links to third party websites are provided by Oxford in good faith and for information only. Oxford disclaims any responsibility for the materials contained in any third party website referenced in this work

General Manager: Laura Pearson
Executive Publishing Manager: Erik Gundersen
Managing Editor: Louisa van Houten
Associate Editor: Yasuko Morisaki
Director, ADP: Susan Sanguily
Design Manager: Lisa Donovan
Senior Designers: Debbie Lofaso, Yin Ling Wong
Electronic Production Manager: Julie Armstrong
Production Artists: Elissa Santos, Julie Sussman-Perez
Image Manager: Trisha Masterson
Image Editor: Liaht Pashayan
Production Coordinator: Chris Espejo

ISBN: 978 0 19 467128 6 STUDENT BOOK (PACK COMPONENT)
ISBN: 978 0 19 467158 3 STUDENT BOOK (PACK)
ISBN: 978 0 19 467198 9 ACCESS CARD (PACK COMPONENT)
ISBN: 978 0 19 467203 0 ONLINE PRACTICE (PACK COMPONENT)

Printed in China

This book is printed on paper from certified and well-managed sources

ACKNOWLEDGEMENTS

*Cover Design:* Molly K. Scanlon

*Illustrations by:* Ben the Illustrator (Agency Rush): pg. 34, 37, 99, 105; Mark Duffin: pg. 31, 32; Seth Erickson: pg. 27, 33, 39 (realia), 45, 61; Alan Kikuchi: pg. 13, 35, 41, 53, 63, 69, 81, 103, 104 (map inset); Chris Lyons: pg. 55, 71, 73, 89 (food insets), 93; Cynthia Malaran: pg. 11, 23, 55, 101; Chris Masterson: pg. 10, 42 (tablet screen inset), 43 (tablet and screen); Stacy Merlin: pg. 5, 17, 51, 57, 67, 73, 79, 85, 89 (breakfast realia), 107, 113; Joe Taylor: pg. 39 (before and after images), 95 (people).

*Photography by:* Ken Karp

*We would also like to thank the following for permission to reproduce the following photographs:*

*Cover photos:* AM-STUDiO/shutterstock.com, olly/shutterstock.com, LosevskyPavel/shutterstock.com, kohszekiat/shutterstock.com, Mika Heittola/shutterstock.com, Konstantin Sutyagin/shutterstock.com, ArtemZhushman/shutterstock.com, ilolab/shutterstock.com, Jason Stitt/shutterstock.com, Nickolya/shutterstock.com, shock/shutterstock.com, Shvaygert Ekaterina/shutterstock.com, AJP/shutterstock.com, BestPhotoStudio/shutterstock.com, Yuri Arcurs/shutterstock.com, JonMilnes/shutterstock.com, qingqing/shutterstock.com, Goodluz/shutterstock.com, Matthew Williams-Ellis/shutterstock.com, BestPhotoStudio/shutterstock.com, Christian Bertrand/Shutterstock.com, javi_indy/Shutterstock.com.

pg. ii Mark Bowden/istockphoto.com; pg. iii Steve Debenport/istockphoto.com; pg. 3 (Argentina, Australia, Thailand) adam.golabek/shutterstock.com, (Brazil, Korea, USA) Oxford University Press, (China, Japan, Mexico, Turkey) Graphi-Ogre/Oxford University Press,(Spain, Portugal) granata111/shutterstck.com; pg. 4 Image Source/Alamy; pg. 5 TroelsGraugaard/istockphoto.com; pg. 7 Still images courtesy of People's Television; pg. 9 (eat) Mike Goldwater/Alamy, (finish) Stockbyte/Getty Images, (get up) PhotosIndia/age fotostock, (bed) rubberball/Getty Images, (go to work) John Lund/Drew Kelly/Getty Images, (tv) Chederros/age fotostock; pg. 11 (jeepney) Victor Fraile/Corbis, (woman) MBI/Alamy; pg. 13 Still images courtesy of BBC Motion Gallery;pg. 15 (gym) moodboard/Alamy, (movie) Indiapicture/Alamy, (piano) Brand X Pictures/Oxford University Press, (soccer) John Kelly/Getty Images, (friends) Daniel Laflor/Getty Images, (yoga) Dani Rodríguez/age fotostock; pg. 16 PhotoTalk/istockphoto.com; pg. 17 Tetra Images/Alamy; pg. 19 Still images courtesy of People's Television; pg. 21 (1) Googolplex/Alamy, (2) Yuri Arcurs/Alamy, (3) Pando Hall/Getty Images, (4) Marcy Maloy/Getty Images, (5) ONOKY - Photononstop/Alamy, (6) Fuse/Getty Images, (7) ACE STOCK LIMITED/Alamy, (8) Westend61 GmbH/Alamy, (9) i love images/Alamy, (Kate) Ken Weingart/Alamy, (10) Kindler, Andreas/Getty Images, (11) Dana Hurse/Masterfile, (12) David Davis Photoproductions RF/Alamy, (13) UpperCut Images/Alamy, (14) Radius Images/Getty Images, (15) Nicole Hill/Getty Images; pg. 22 mauritius images/AGE fotostock; pg. 23 (car) Blend Images/Alamy, (kitchen) Ted Foxx/Alamy; pg. 25 Still images courtesy of People's Television; pg. 27 travel Photodisc/Oxford University Press, (exercise) Catchlight Visual Services/Alamy, (soccer) RubberBall/Alamy,(yoga) Comstock/Getty Images; pg. 29 (beach) Hugo Yonzon/AGE fotostock, (businessman) GlowImages/Alamy, (teammates) Jon Feingersh/Blend Images/Corbis; pg. 33 (brownstones) Walter Bibikow/AGEfotostock, (restaurant) Jean Heguy/AGEfotostock, (stoops) Jupiterimages/Getty Images; pg. 35 Still image courtesy of BBC Motion Gallery; pg. 41 Still images courtesy of BBC Motion Gallery; pg. 45 (barbecue) imagebroker.net/SuperStock, (soccer) Stockbyte/Getty Images; pg. 47 Still image courtesy of People's Television, (b/w photo) Emiliano Ortiz;pg. 49 (chef) Tetra Images/Oxford University Press, (clerk) Tetra Images/Alamy, (doctor) Image Source/Oxford University Press, (flight attendant) Digital Vision/Oxford University Press, (mechanic) Valueline/Oxford University Press, (server) Leigh Schindler/istockphoto.com; pg. 51 (driver) Cultura/Zero Creatives/Getty Images, (agent) Flying Colours Ltd/Getty Images; pg. 53 Still images courtesy of BBC Motion Gallery; pg. 59 (concert) Tim Tadder/Corbis, (gallery) Axiom Photographic Limited/SuperStock, (movies) Belinda Images/SuperStock, (theater) Mark Wilkinson/Alamy, (shopping) OleksiyMaksymenko/Alamy, (soccer) allfive/Alamy; pg. 60 Jose Fuste Raga/Corbis; pg. 61 (city hotel) Design Pics Inc. - RM Content/Alamy, (marina) nagelestock.com/Alamy; pg. 63 Still image courtesy of BBC Motion Gallery; pg. 65 (art) Ian Shaw/Alamy, (gym) Chris Clinton/Getty Images, (history) PhotoAlto/Sigrid Olsson/Getty Images, (language) Blend Images/Alamy, (math) Asia Images/Getty Images, (science) Image Source/Alamy; pg. 67 Atlantide S.N.C./AGEfotostock; pg. 69 Still image courtesy of People's Television; pg. 72 GlowImages/Alamy; pg. 74 (pharmacy) Jeff Greenberg/AGEfotostock; pg. 75 (video) Still images courtesy of BBC Motion Gallery, (clown) Jose Luis PelaezInc/Blend Images/Corbis; pg. 77 (backpack) HeideBenser/Corbis, (business) Flirt/SuperStock, (camping) Bill Stevenson/AGEfotostock, (cruise) Prisma/SuperStock, (passports) Nicholas Burningham/Alamy, (suitcases) Peter Dazeley/Getty Images; pg. 79 (Antonio) Jupiterimages/Getty Images, (Hyun) Luca DiCecco/Alamy, (Sandy) Janine WiedelPhotolibrary/Alamy; pg. 81 Still image courtesy of BBC Motion Gallery; pg. 83 (Barcelona) Mike Randolph/Masterfile, (lake) Robert Harding Picture Library/SuperStock, (London) Jan Tadeusz/Oxford University Press; pg. 85 vm/istockphoto.com; pg. 87 (1) Digital Vision/Oxford University Press, (2) MIXA/Oxford University Press, (3) Good Shoot/Oxford University Press, (4, 5) Stockbyte/Oxford University Press, (6) Lothar Wels/Masterfile, (7, 8, 12) Photodisc/Oxford University Press, (9) EYESITE/Oxford University Press, (10) Photocuisine/Masterfile, (11) Fancy/Oxford University Press; pg. 91 Still images courtesy of People's Television; pg. 97 (saris) Martin Harvey/Getty Images, (video) Still image courtesy of BBC Motion Gallery; pg. 99 (cloudy) Marka/SuperStock, (foggy) Michael DeYoung/AGEfotostock, (rainy) Adriano Schena/AGEfotostock, (snowy) Gary Gerovac/Masterfile, (sunny) RaimundLinke/Masterfile, (windy) PATRICK LIN/AFP/Getty Images; pg. 101 (compass) Akai37/shutterstock.com, (woman) Thomas Barwick/Getty Images; pg. 103 Still image courtesy of BBC Motion Gallery; pg. 107 Russell Monk/Masterfile; pg. 109 (Flatiron) JorgHackemann/Shutterstock.com, (video) Still image courtesy of People's Television; pg. 111 (China) Steve Vidler/AGEfotostock, (Egypt) sculpies/shutterstock.com, (Brazil) OSTILL/istockphoto.com; pg. 113 (Canada) Walter Bibikow/AGE fotostock, (woman) Christopher Oates/shutterstock.com.

*Additional photography provided by:* SergejKhakimullin/shutterstock.com, Anna Maniowska/istockphoto.com, Rubberball/istockphoto.com, peter zelei/istockphoto.com, Tom Young/istockphoto.com, SupriSuharjoto/shutterstock.com, Tom Wang/shutterstock.com, Kurhan/shutterstock.com, Felix Mizioznikov/shutterstock.com, KaterynaUpit/shutterstock.co, Marko Tomicic/shutterstock.com, James Peragine/shutterstock.com, Christopher Oates/shutterstock.com, AISPIX/shutterstock.com, Matthew Williams-Ellis/shutterstock.com, Peter Kirillov/shutterstock.com, AndreyArkusha/shutterstock.com.

Video produced by: People's Television, Inc. www.ppls.tv